Dynamics of Land Markets and Emerging Land-Use Pattern

About the Authors

Prem Singh (Dr.), IAS [Former Centre Director, Centre for Rural Studies and Deputy Director (Sr.), LBS NAA, Mussoorie], **Batch:** 2000, **Cadre:** Manipur-Tripura, MBBS from JIPMER, Puducherry. Worked as SDO Chandel; Director, Youth Affairs and Sports, Manipur; Deputy Commissioner, Senapati District; Secretary to CM and Director, Municipal Administration Housing and Urban Development. He has done Post Graduate Program in Public Management and Policy (PGP-PMP) from IIM, Ahemadabad. Was posted as Director, Municipal Administration Housing and Urban Development, State Mission Director (NRHM) and State Project Director (SSA). Currently he is posted as Director (Japan), Department of Economic Affairs, Government of India, Ministry of Finance, New Delhi. Areas of interest include applied economics of welfare state, service delivery in social sector (health, education).

Dr. Varunendra Vikram Singh, ***Faculty, Centre for Rural Studies:*** Ph.D., Sociology Department, Lucknow University, Lucknow. Awarded Junior Research Fellowship and also Senior Research Fellowship by Anthropological Survey of India, Government of India, Kolkata. Did Certificate Course in 'Computer Basics & Internet'. Passed National Stock Exchange's Certification in 'Financial Markets with expertise in Derivative & Capital Market.' Did an International Course on Management of 'Common Property Resources' from IIM, Ahmedabad. Organized and contributed to several seminars and workshops of national importance. Worked as Researcher in Nationa' Institute of Administrative Research, LBS National Academy of Administration, Mussoorie. Has several Reports and Research papers published in scientific journals and also edited published and Co-authored books to his credit. Areas of Interest are in the field of Land Reforms, Land Markets, Elementary Education, Tribal Development, Common Property Resources, Financial Management and Land Resources Management in India.

Dynamics of Land Markets
and
Emerging Land-Use Pattern

Dr. Prem Singh, ***IAS***
Dr. Varunendra Vikram Singh

Published for

LBSNAA
Lal Bahadur Shastri National Academy of Administration

Centre for Rural Studies
Lal Bahadur Shastri National Academy of Administration
Mussoorie
by

CONCEPT PUBLISHING COMPANY PVT. LTD.
NEW DELHI-110059

ISBN-13: 978-93-5125-248-1

First Published 2016

Published and Printed by

Since 1974 **Concept Publishing Company Pvt. Ltd.**
Regd. Office:
A/15-16, Commercial Block, Mohan Garden,
New Delhi-110059 (INDIA)
Phones : 25351460, 25351794, *Fax* : 091-11-25357109
Email : publishing@conceptpub.com
Website: www.conceptpub.com

Editorial Office: H-13, Bali Nagar, New Delhi-110 015, India.

Cataloging in Publication Data—DK Courtesy: D.K. Agencies (P) Ltd. <docinfo@dkagencies.com>

Singh, Prem, 1973- **author.**
Dynamics of land markets and emerging land-use pattern / Dr. Prem Singh (IAS), Dr. Varunendra Vikram Singh.
pages cm
Includes bibliographical references and index.
ISBN 9789351252481

1. Land use—India—Dehra Dun (District) 2. Land tenure—India—Dehra Dun (District) 3. Real property—Valuation—India—Dehra Dun (District) I. Singh, Varunendra V. (Varunendra Vikram), author. II. Centre for Rural Studies (Lal Bahadur Shastry National Academy of Administration) III. Title.

HD880.D44S56 2016 DDC 333.7313095451 23

Preface

This book is a narrative exposition of emerging issues in the dynamics of land markets in the background of economic development in India. Progressively land is emerging as a market commodity in India and land and labor are the basic and fundamental resources available to the people for income generation. In this context the fast changing and emerging land use pattern has improved the access to land and employment for marginal and landless people and has also became basic determinants of well-being for the rural areas. Land markets are habitually seen as unfriendly to rural areas and the poor, although they may be beneficial not only for overall growth, but also for rural growth and poverty alleviation.

Land markets are leading to less landlessness rather than more and functioning of land markets is basically look at the land delivery system from two angles i.e. land sales markets and land rental markets. Land rental markets are leading to transferring land to more productive users at a very low cost. Whereas the land sales activity is related to permanent freehold and long term leases mainly set off for non- farm activities and also it accelerate the greater crop diversification, higher yield and livestock revenues. Under land market activities labor employment is showing diversity of occupations but the prime reliance on farm land is still persists. Land markets also enhance non-farm employment which is associated with higher overall employment. Largely, land markets do not show adverse consequences but rather some positive impact on rural and urban land.

We hope this book will provide valuable insight on the various existing dimensions of land markets in a wide-ranging socio-economic context.

Dr. Prem Singh
Dr. Varunendra Vikram Singh

Acknowledgements

While conducting the study and writing the report I received help in one or the other way from various persons, so I am gratefully thanking them all at this available opportunity.

First and foremost I am grateful to my Centre Director, Dr. Prem Singh, IAS, of Centre for Rural Studies, Lal Bahadur Shastri National Academy of Administration, Mussoorie for extending approval of the study to me with encouragement and able guidance. Despite being very busy his sincere guidance at every step has made the report meaningful. It is my pleasing duty to thank all the faculty members of Centre for Rural Studies for cooperation and thanks are due to staff of Centre for Rural Studies, who accepted the task of data entry and secretarial assistance.

Thanks are also due to Sri B.S.R.V. Prasad, District Magistrate, Dehradun and Sri Harak Singh Rawat, Additional District Magistrate, Dehradun along with SDMs of three Tehsils namely Dehradun, Vikasnagar and Rishikesh respectively for providing administrative help during field work. I am equally thankful to Tehsildars of Dehradun, Vikasnagar and Rishikesh, Tehsils for providing necessary insight in the subject and region. I am also indebted to all Patwaries/Lekhpals of these three Tehsils for supporting and conducting the data collection work for the study and help in the field work. Special Thanks to Ms. Rita Sinha, IAS (Retd.), Shri Charanjit Singh, IFS and Dr. D.Sajith Babu for editing the final report.

In the last but not least all the prominent informants hailing from the studied villages deserve thanks for easing my field work and study.

Dr. Prem Singh
Dr. Varunendra Vikram Singh

Contents

Preface v
Acknowledgements vii

1. Theoretical Proposition 1
2. Introduction of the Study Area 7
3. Dynamics of Land-Use: Agriculture and Allied Sector Economy 16
4. Dynamics of Land-Use: Economy from Forest Resources 23
5. Dynamics of Land Markets: Emerging Issues and Nature of Functioning 28
6. Status of Land Markets, Productivity and Magnitude 41
7. Concluding Remarks 55

Appendices 60
Bibliography 85
Index 88

CHAPTER 1
Theoretical Proposition

Conceptual Background

All activities of the living world are basically centred on Land Surface of earth, as our existence is based mostly on land. Land is the source of all socio-economic activities connected with food, shelter and material culture. Primary, secondary and tertiary economic activities are key to the development of a society and all these activities are connected directly or indirectly with the land.

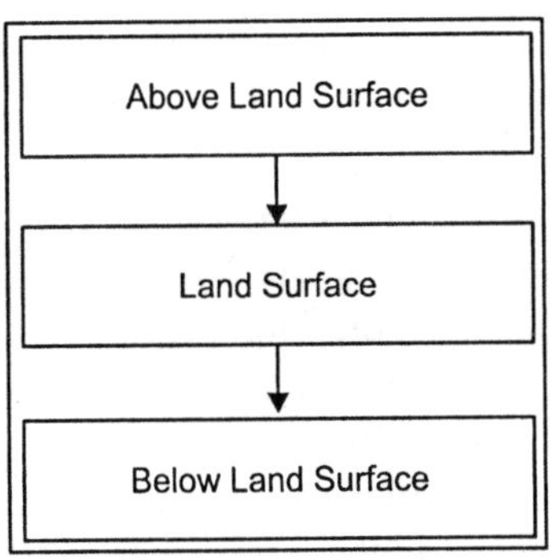

Fig. 1.1: Land in three Dimensional Resources

In economic terms, land can be seen as a three-dimensional resource as shown in Figure 1.1, which is an important source of economic resources. Communities and governments may have ownership issues regarding control and extent of use of it, and land being an immovable resource, during land transactions only ownership rights is transferred. Land always remains static but land-use of the land is changed or replaced. Land rights and its use are always subjected to conveyance but land remained physically attached to the surface of earth and politically with the governments. It cannot be moved under any circumstances. The above land transactions which results in exchange of rights and ownership are called "Land Markets". Money or capital is the key factor to motivate the "land market" activity. In Indian context and particularly in the study area of district Dehradun, "land markets" have been motivating the industrialization and real estate substantially. As per Robert Mahaoney (2007) for developing a " 'land market' a number of critical factors need to be in place. In addition to clear policies and enforceable laws the institutional framework needs to provide for consistent responsible and accountable decision-making within the overall framework of National land policy. There must also be a land registration system that is affordable so that all citizens especially women and minority groups, rich and poor, can have access to it".

"Land markets" itself decide the ownership and price mechanism of the market. For more efficient use of "land market" it should be developed into a more transparent, effective land administration system and in an eco-friendly manner.

"Actors in the market are usually only active infrequently during their lifetimes. Land transactions are however after the largest single transactions made by an individual and require specialist knowledge. It is, therefore, common practice for vendor and perhaps the purchaser to use specialist advisers to assist in the transaction". (Foncier, 1999)

"Land Market" is a place where land sellers and purchaser exchange their rights and interests on land. In the "Land Market" institutional framework, social practices, relationships and regulatory mechanisms take place under consideration. These factors become relevant/irrelevant as per the supply and demand of the market players. It is important to note that whether the market is working or functioning formally or informally is beyond any consideration. "Land Markets" are also taking their shape by the

State government policies as well. Kihato and Momoniat (2010) are of the opinion that "these are enforced by a variety of bodies, including the state, consumer rights organizations, international trade organizations and localized community or social organizations. Depending on how sophisticated the market is the number of these components and their relationships can vary considerably. It is important to note that all markets operate within a broader political economy where the power balance between the different actors/institutions is unequal".

Vermeulen (2009) has advocated that "great majority of the world's population and all those living in developed economics, live in societies in which the goods and services they consume are provided through markets and subject to their incomes—they are free to choose what, where and how much to consume. But to varying degrees governments interview and regulate all markets. So, the regulation of "land markets" is not exceptional in itself but it is exceptional in its form and severity."

The functions of "land markets" are centred on the transfer of ownership rights or property rights. As per Birner (1999) they include the right to use, to manage, to generate income to temporarily or permanently transfer the land (including to lend on mortgage) the right to exclude others and the right to compensation. Finally, complementary institution in the form of financial mechanisms can improve the functioning of "land markets" and such on economically efficient "land market" will however have a number of negative social and environmental side-effects. (Wehrmann 2006 and Babettee, 2008).

The transactions of "land market" are an old phenomenon but the dynamics of "land markets" study is very recent one. However, many eminent scholars have studied several important aspects of "land market" from different parts of the world but unfortunately no one from the study area, i.e., District of Dehradun. Among these mention may be made of Peter Dale, R.M. ("land markets" and the Modern economy); Simpson, S.R. (Land, Law and Registration); Kivell, P.T. (Public Ownership of Urban Land); FAO (Land Tenure and Rural Development); De Soto, H. (The Mystery of Capital); Feder F. *et al.* (Land Policies and Farm Productivity in Thailand).

The "land market" in Indian context, works in two paradigms as indicated in Figure 1.2:

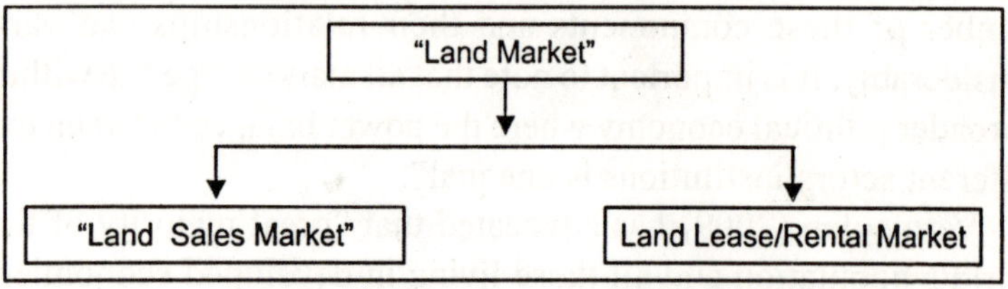

Fig. 1.2: Land Market in Indian Context

Land sales market facilitates transfer of land through permanent sale by relinquishing all the rights of the seller. Usually, land in this type of transaction goes in the non-agriculture sector and owners are generally are the settlers. In land rental or lease markets emphasis is given on share tenancy and contracts. Land rental markets are the means of achieving optimal size of functional land holdings with full utilization and employment of family labour.

During 1950s land reforms was adopted as on agenda of development by abolition of intermediaries, tenancy and land ceiling provisions were introduced to curtail amassment of land masses by a few. By 1970s, rural poverty was substantially reduced owing to land reforms and abolishment of intermediaries and it became relevant in the context of development. Now during last few years "land markets" are appearing as an important ingredient for economic development of the communities. There is a need to look "land market" as a supporting factor in a broader strategy of area development where state should provide secure and unconditional rights over lands and there should be an equal opportunity market which is techno-friendly and supportive to all land owners including small and marginal land owners.

Uttarakhand became an independent State in the year 2000. It has a geographical area of 20,682 square miles (53,566 km²). The largest cities in the State include Dehradun (530,263), Haridwar (220,767), Haldwani (158,896), Roorkee (115,278) and Rudrapur

(88,720). The State government recognizes 15,620 villages and 81 cities and urban areas.

Since, inhabitants of the study area, i.e., district of Dehradun are venturing into new horizons of development, over and above the subsistence limits on food, water and dwelling, the activities whether it is for meeting basic needs or for economic growth dependent on the land. How should land be made available for "Land Market" and its optimum utilisation in favour of the land owners to make it useful for the benefit of all socio-economic groups of the district Dehradun, has been a crucial question. It needs to develop a policy by the Government of Uttarakhand which should own land for non-agricultural purposes also maintain a regulatory framework for its effective management. It is a fact that "land market" are directly functioning at the doorstep of every land owners and during every land transactions it is important to find out the ways and means by it can be effectively monitored to maintain the effectiveness and equilibrium for the development of "land market".

To study the various existing dimensions of "land market" in district Dehradun and to develop a community owned and managed system of "land market" in wider socio-economic context a monitoring mechanism is highly necessary. In this backdrop the present study has been carried out with the following objectives:

1. To fix the study area to get maximum reflection of the subject under study.
2. To assess the dynamics of land-use with respect to economy of agricultural and allied sector.
3. To assess the dynamics of land-use with respect to the contribution of forest resources in the economy.
4. To study the dynamics of "land market" at the backdrop of emerging issues and nature of functioning.
5. To study the State's, productivity and magnitude of "land markets" in the district.

Design of Research and Methodology

For undertaking the detailed and intensive study of the valley region, i.e., three "land market" active Tehsils of Dehradun district, viz., Vikasnagar, Dehradun and Rishikesh were selected. Since, altitude is a determining factor for forming agro-climatic zones in the Himalayan set-up, selection of villages for the study was done in a scientific manner to make the sample adequately representative.

For collecting primary data the fieldwork was carried out amidst all issues inherent to the study area. The data was collected by convincing the people after establishing good rapport with them. The data were collected by administering standard empirical methods and techniques used in social sciences such as schedules, interview, etc., containing a list of questions logically related to the theme of study. Participant observers collected the data through keen observation of the activities, directly and indirectly connected to "land market". The information was further corroborated through several village visits. During the field visits hard copy documents were collected from various government offices in the form of monographs, texts and Gazetteers, etc.

Both quantitative and qualitative tools of analysis were used in the study to analyze the status, role and utilization of "land market" and institutional management in the traditional and modern framework and its related issues. Accordingly, four schedules were designed to collect the data as given in Appendix B.

The primary data was collected through extensive primary survey and compiled with secondary data from various government departments. Both primary and secondary data collected by different methods were subjected to analysis on its interactions, inter- and intra-dependence. The idea was to comprehensively interpret the data with reference to prevalence and relevance of "land market" in the study area with a view to correlate the same with the theoretical orientation outlined in the study.

Chapter 2
Introduction of the Study Area

District of Dehradun is situated in the Western region of the State of Uttarakhand, having an area of 3,088 sq. km., spread between 29°-58′ to 31°-2′ North latitudes and 77°-35′ to 78°-18′ East longitude. On the north, it is bounded by the district Uttarkashi, on the east by Tehri district and on its west lie the neighbouring State of Himachal Pradesh. The river Yamuna and the Ganges comprises its western and eastern boundaries respectively. Dehradun can be divided into two distinct tracts, i.e., the montane tract and the sub-mountain tract. The mountain tract covers whole Chakrata tehsil of the district and consists entirely of a succession of mountains and gorges and comprises Jaunsar-Bhabar. The mountains are very rough with steep slopes. Below the mountain tract is the sub-mountain track, which is famous Dun valley bounded by Shivalik hills in the south with outer scrap of the Himalayas in the north.

Physiography

The entire area of Dehradun district consists of array hills, valleys and mountains with a range of elevation from about 600 m to 3000 m above Mean Sea Level (MSL). It is a complex mosaic of hills, valleys and snow-clad peaks. The steep mountain slopes are covered with coniferous and broad leaved trees, meadows, pasture lands and terraces, often clinging precariously on the mountain sides. The valleys lying amidst these mountain ranges contain gurgling streams, torrents, terraced fields and fairy tale villages.

The hills of this region may be subdivided into the following physiographic units (from south to north):

(1) Outer Himalaya or Shiwalik Hills.
(2) Lower or Lesser Himalaya.
(3) Main or Greater Himalaya.

District of Dehradun has a complex geomorphology, lithology, structure and stratigraphic succession. The geology of this region resembles that of Himachal Pradesh in the west and Kumaon in the east.

Drainage

District of Dehradun is extremely rich in hydrological resources, viz., water and snow. The whole region lies within the catchment of the Yamuna and Ganges rivers. The central ridge stretching from south-west to north-east direction in the centre separates these two systems.

Climate

The climate of District varies from sub-tropical in the southern tract to sub-arctic in the snow-bound areas of the north. A large part of this region is influenced by the south-west monsoon rains.

Climatic Regions of District of Dehradun

The following climatic regions have been delineated in Dehradun area:

(1) ***Sub-tropical Region***: This climatic region includes all areas lying below an elevation of about 800 m above MSL usually encompassing parts of the Shiwalik hills, Bhabar tract and the lower Himalaya. The climate of this region is largely free from extremes of weather. The mean annual temperature varies from 18°C to 29°C with the maximum

temperature rising to more than 38°C at times. In winter the mercury may drop to near freezing points and heavy rainfall is received in this region during the south-west monsoon. Area seldom experiences snowfall.

(2) ***Sub-temperate Region*:** This climatic region comprises of area lying between an elevation of 800 and 2000 m above MSL, covering part of the lower and middle Himalaya. The climate of this region is largely free from extremes of hot and cold conditions. The mean annual temperature ranges from 14°C to 19°C with the maximum summer temperature in the vicinity of 35°C. This region experiences cool winter season. Snowfall is received in the upper parts of the sub-temperate region, though snow does not sustain for more than two or three days. The south-west monsoons results in heavy and wide spread rains in this region.

(3) ***Temperate Region*:** This region cover areas lying between an elevation of 2000 and 2400 m above MSL, in the middle of higher Himalaya of district Dehradun. The mean annual temperature varies from 11°C to 15°C with the mercury soaring to over 25°C during summer. The average annual rainfall in this tract varies from 80 to 200 cm. Bulk of the total annual precipitation is received from the south-west monsoon. Winters are fairly cold in this region and heavy snowfall experienced during this season. The snow may cover the ground for several days at a stretch.

(4) ***Sub-Arctic Region*:** This region covers areas lying between an elevation of 2400 and 3000 to 3200 m above MSL in the middle and higher Himalayas of Dehradun district. The mean annual temperature ranges from 8°C to 12°C. In summer the mercury usually does not rise above 22°C. The average annual rainfall is relatively less than that received in the lower elevations. Winters are very cold and severe. The temperature drops down even below the freezing points and there occurs heavy snowfall and thick snow cover the region for long periods throughout the winter season.

Cycle of Seasons

District of Dehradun experiences the following well-defined cycle of climatic seasons:

(i) ***Summer*:** The summer season in most parts of district of Dehradun begins from the middle of April and continues till late June till the south-west monsoon arrives. However, the beginning of the summer season may be late in the high altitude areas. The end of this season may be marked by pre-monsoon showers. It becomes extremely hot in the lower Himalaya and Valleys of the higher Himalaya in the summer season. During June the mercury may rise to more than 35° in the southern parts. The areas lying at higher elevations are fairly cool even in summer.

(ii) ***Monsoon*:** The south-west monsoon causes heavy and widespread rains in most parts of District of Dehradun from the end of June to the middle of September. The monsoon rains brings relief to this tract which is reeling under an intense spell of heat in the summer season. The rains are very heavy on the south facing slopes of the mountains which may continue for several days at a stretch in many areas. Vast areas are covered with mist for long periods and the sun rarely comes out. Widespread havoc is caused by the monsoon rains as it results in heavy damage of roads and bridges through flash floods and frequent landslides usually disrupt normal life. Loss of human life and property are common phenomenon during this period.

(iii) ***Autumn*:** The monsoon rains stop in the middle of September or early October. The weather clears up and bright sunny days bring relief from the prolonged spell of rains. The weather remains excellent in most parts of this mountain chain, except at very high elevations where winter quickly sets in and the first snowfall of the season may occur in late September. Autumn continues till the end of November after which the long Himalayan winter sets in.

(iv) ***Winter***: The winter seasons of this region is fairly long and dull. It begins in late November in the upper reaches and may continue till the middle or end of March. At lower elevation the beginning of the winter seasons may be delayed till the middle of December. Snowfall is common in most tracts of upper regions of Dehradun district lying above an elevation of about 2300 m above MSL. The snow may remain the covering the ground for several weeks at a stretch. The mercury remains below the freezing points in many parts of upper Dehradun region. Frost is experienced in winter in the valleys of rivers and streams. Late winter frost may occur as late as in mid-March.

(v) ***Spring***: The season brings relief from the long and dull winter season. Spring begins from the middle of March and continues till the arrival of the summer season in mid or late April. This season marks the transition from winter to summer. Late night frost may be experienced in the early part of the spring season.

Rainfall and Snowfall

Bulk of the total annual precipitation in most parts of Dehradun region is received in the form of rain. The amount and intensity of rain is influenced by the location with respect to the advancing monsoons and convection of air currents. The average rainfall received at Chakrata Township is 155.7 cm per year. Areas lying above an elevation of about 2000 m above MSL receive snowfall during the winter months. While snowfall may occur only a few times in areas lying between 200 and 2500 m above MSL and its frequency is much more at higher elevations.

Soil

The soil is the basic necessity for the agriculture. In the district of Dehradun soils with moderate fertility occur in the Yamuna and Ganges valleys. Soils in the upper valleys are by and large fertile due to the present of a large quantity of humus, minerals and moisture. The alluvial soils of the valley bottom are more fertile.

Natural Vegetation

The Himalayan mountain chain in Dehradun region bears rich forests, with thousands of different species of trees, shrubs, herbs and climbers. The nature of forests found in a particular locality depends upon the prevailing climatic conditions, altitude, aspect, topography, soil and biotic influences. In Dehradun region the following forests groups have been recognized:

(i) ***Sub-Tropical Forest Region***: This forest region encompasses the southern tract of Dehradun region. The altitudinal range is from 250 to about 1500 m above MSL and the principal forest types of the forest region are:
   (a) Sal forests.
   (b) Riverine Khair and Shisham forests.
   (c) Chir (Pine) forests dominated by *Pinus roxburghii.*

(ii) ***Temperate Forest Region***: Temperate region covers the tract between an elevation of about 1500 m to 3500 m above MSL. It mainly comprises of Deodar (*Cedrus deodara*), Oak, Kail (*Pinus wallichiana*), fir (*Abies pindrow*), spruce (*Picea smithiana*) and mixed broadleaved forests trees consisting of *Aesculus indica, Rododendron arboreum, Juglans regia*, etc.

(iii) ***Sub-Alpine and Alpine Region***: This forest region encompasses the upper most limit of tree growth, usually extending from an elevation of about 3000 m above MSL up to the snow line. In this sub-alpine and alpine zone the main trees are alder (*Alunus nitida*), fir (*Abies spectabilis*), kharsu, birch, chestnut, walnut, apricot, etc. The treeless slopes are covered with moist alpine shrubs along with alpine grasses.

Fauna

From the dense Sal forest of the lower heights to the higher elevations, Dehradun region is the home of thousands of interesting species of mammals, birds, reptiles and amphibians. North of the main Himalayan chain lies the Palearctic region. The fauna of this Himalayan region resembles that of both the oriental region as well as that of the Palearctic region. The Dehradun region which considered rich in wild animals seldom exhibits the diversity now. However Uner (tiger), Rikh (Bear), Siyad (Jackal), Wolf, Jodav (Hog Deer), Cheetal/Ther (Spotted Deer), Kothur (Musk Deer), Sungur (Pig), Gurat, Kakad (Barking Deer), Kursiyalti, Kuni, Shail/ Sehi (Percupine), Bander (Monkey), Langoor (Black Monkey), Ghoral etc., mammals; Chiod, Vank Kukde, Titeri, Gada, Chillad, Malkhery, Mor, Gugtia, Udar, Kothodar, Thoka, Phenjua, Devchadi, Kumarchodo, Tilbuchh, etc., dominates among birds and Supin, Chhatwah, Senvadia, Oraltu, Bodai, Didothia, Dudhia, etc., are the snakes found in Mountainous region of the district.

Population and Literacy

The settlement pattern in the district is distinctively of two types. The villages in mountainous tract are small in size and population is mostly scattered, whereas in valleys, villages are with dense population. As per the 2011 Census of India the Dehradun district has a population of 16, 98,560 the second highest in Uttarakhand after Haridwar (19, 27,029). The Decadal growth rate has increased from 25 per cent (1991-2001) to 32.48 per cent (2001-2011). This increase is the third highest in Uttarakhand after Haridwar (33.16%) and Udham Singh Nagar (33.40%). The district Dehradun has a gender ratio of 902 as against a State average of 963. The population density is 550 persons per sq.km. again the third highest after Haridwar (817) and Udham Singh Nagar (648). The State average is 189 persons per sq.km. The literacy rate is the highest in the State at 85.24 per cent (90.32% for males, 79.61% for females) of literate district.

According to the 2011 census, Dehradun district has a population of 1,698,560, roughly equal to the population of Guinea Bissau or the US State of Idaho.

As per the sample survey conducted in 4 blocks of the district of Dehradun, it is understood that Sahaspur block has the highest number of around 28.0 per cent educated persons, second is Doiwala (26.0%) and Vikasnagar was the lowest at 21.0 per cent. Doiwala block has shown highest number of persons who are educated up to intermediate, graduate, post-graduate and technical/ professional level. Overall inference is that, people in the region are competent and educated to get employed in the "land markets". The details are in Table 2.1 below.

Table 2.1: Percentage of Educated Population

Block	*Primary*	*High School*	*Inter-mediate*	*Graduate/ Post Graduate*	*Technical Graduate*	*Total*
Doiwala	84 17.6%	194 24.9%	92 30.7%	91 42.7%	5 45.5%	466 26.0%
Raipur	119 24.9%	193 24.8%	79 26.3%	47 22.1%	1 9.1%	439 25.0%
Sahaspur	124 25.9%	244 31.3%	86 28.7%	46 21.6%	2 18.2%	502 28.0%
Vikasnagar	151 31.6%	148 19.0%	43 14.3%	29 13.6%	3 27.3%	374 21.0%
Total	478 100%	779 100%	300 100%	213 100%	11 100%	1781 100%

Source: Field Survey, 2014-15.

It is also observed in the sample survey that Vikasnagar block has the highest number of Scheduled tribe and Other backward class population. Block Sahaspur is housing the highest number of Scheduled caste population. Whereas, in Raipur block highest number of families from general category found to inhabit. It was observed that the "land market" was more actively functioning in the Vikasnagar and Sahaspur blocks where the predominant categories were the SC, ST and OBCs. In Raipur block where the families from the general category dominated, the "land market" was comparatively dull. The details are given in Table 2.2.

Table 2.2: Caste-wise Percentage of Population

Block	*General*	*OBC*	*SC*	*ST*	*Total*
Doiwala	342 26.90%	101 21.70%	40 14.60%	2 2.40%	485 23.16%
Raipur	428 33.70%	53 11.40%	35 12.80%	0 0.00%	516 24.64%
Sahaspur	358 28.20%	118 25.30%	100 36.50%	6 7.20%	582 27.79%
Vikasnagar	143 11.30%	194 41.60%	99 36.10%	75 90.40%	511 24.40%
Total	1271 100%	466 100%	274 100%	83 100%	2094 100%

Source: Field Survey, 2014-15.

As per census 2011, there are around 902 females to thousand males in the district of Dehradun. The disaggregated data from the sample survey in the four blocks of Doiwala, Raipur, Sahaspur and Vikasnagar revealed that sex-ratio in adjacent urban blocks and the blocks located far from Dehradun Urban Agglomeration. Sex ratio is relatively better in comparison to the overall district sex ratios and the details are given in the Table 2.3.

Table 2.3: Male-Female ratio of Families

Block	*Male*	*Female*	*Children*	*Total*
Doiwala	1222 23.00%	1093 23.00%	989 20.60%	3304 22.24%
Raipur	1247 23.50%	1166 24.50%	1075 22.40%	3488 23.48%
Sahaspur	1360 25.60%	1232 25.90%	1193 24.90%	3785 25.47%
Vikasnagar	1477 27.80%	1262 26.60%	1542 32.10%	4281 28.81%
Total	5306 100%	4753 100%	4799 100%	14858 100%

Source: Field Survey, 2014-15.

CHAPTER 3
Dynamics of Land-Use
Agriculture and Allied Sector Economy

Dehradun district has more diversified and urbanized economy than the other hilly districts of Uttarakhand. However, the economy of the district have two distinct spatial variants, viz., the economy of the mountain tract of Jaunsar-Bawar (Chakarata Tahsil) which is predominantly tribal and traditionally agricultural and of the valley segment with relatively better infrastructure and high rate of urbanization, due to highly active "land market" transactions.

The inhabitants of district of Dehradun primarily depend on land for food production. The hilly people are basically engaged in agriculture and animal rearing which are supplemented by gathering, hunting and fishing from common property resources. The mountain slopes, tops, river banks and valleys are exploited as arable land. In Dehradun valley which is spread from Vikasnagar to Rishikesh, people are engaged in service sector, small scale industries, dairy and business activities for their livelihood.

As per the sample survey given in the following Table 3.1, it is revealed that most of the people in the blocks of Doiwala, Raipur, Sahaspur and Vikasnagar fall under the income group of less than Rs. 5,000 per annum. It shows that, most of them are engaged in primary sector activities (agriculture, horticulture, mining, and forestry) and business activities and "land market" employment are not being taken up seriously.

Table 3.1: Annual Income from Service

(In Rupees)

Block	*Less then 1000*	*1001-2000*	*2001 - 5000*	*5001 - 10000*	*10001 - 30000*	*Above 30000*	*Total*
Doiwala	320 22.3%	108 41.4%	59 25.0%	1 1.1%	0 0.0%	0 0.0%	488 23.2%
Raipur	410 28.6%	37 14.2%	62 26.3%	6 6.7%	1 1.2%	1 100.0%	517 24.6%
Sahaspur	391 27.3%	76 29.1%	37 15.7%	69 77.5%	10 12.2%	0 0.0%	583 27.7%
Vikasnagar	312 21.8%	40 15.3%	78 33.1%	13 14.6%	71 86.6%	0 0.0%	514 24.5%
Total	1433 100.0%	261 100.0%	236 100.0%	89 100.0%	82 100.0%	1 100.0%	2102 100.0%

Source: Field Survey, 2014-15.

The arable land as private property resource is owned and inherited by the individuals, whereas the rest of resources are treated either as Panchayat or State property and have only selectively common access and use right, but without any right of inheritance. Agriculture is fully developed in the region. The entire agricultural operations and cropping pattern, prevalent in the villages selected for study is clearly representative of the region. The harvesting is done manually by using *daranti* (sickle) and threshing is done mostly in courtyard of the house while in valley regions modernized agricultural practices are also prevalent.

The food production systems being an integral part of social sustenance and it integrates different people in the society, who with their various skilled labour contribution, earn the share by taking part in the production-distribution chain and thus meet their annual foodgrain requirement. This practice over centuries resulted in division of labour and social order in tune with it and the systems still remains fruitfully intact and operative.

Besides agriculture, there are a few artisans groups who are from different caste sub-groups. Among this mention may be made of carpentry, tailoring, masonry, smithy, basketry and leather works, etc. Poultry herding and horticulture also contributes to the food production arena.

Agriculture

The basic requirement of agriculture in the district is plain land with fertile soil. In valley region the land is basically plain whereas the highland mountains of the region, very little naturally levelled land can be seen. Therefore, for easy cultivation and checking of soil erosion, terraced fields are prepared. The practice of terrace cultivation in mountainous region of district, have become an integral part of the monsoon zone, for sustainable agriculture. The irrigated terraces are locally known as *kyari*. These are usually located near the valley bottom along the edges of rivers and streams and sometimes these irrigated terraces are also found near the ridge. While those on upland slopes at high altitudes lacking irrigation facility are sown with dry crops or those varieties of crops which need little water. These terraces are locally known as *okhad*.

Agriculture in this area is affected by three important factors: the composition of the soil, the position of the field that is to say its height above sea level and its aspect or lie and the irrigation facilities. In all Dun Valley blocks irrigation facilities are better as 85 per cent of the net sown area is with irrigation facility. In Doiwala block irrigation facilities are well developed, as in the blocks mentioned earlier. Because of ensured irrigation facilities in the valley, consumption of fertilizer and mechanization of agriculture operations has been substantially increased.

Table 3.2: Type of Agricultural Land—Block: Doiwala

Sl. No.	*Village*	*Agriculture Land (Area)*	*Irrigated Land (Area)*	*Unirrigated land (Area)*
1.	Markham Grant II	674	631	43
2.	Laxiwala	10.426	10.426	0
3.	Dasswala	26	25	1
4.	Ghirpari	29.061	28.613	0.448
5.	Hansawala	72	70	2
6.	Doiwala	50	47	3
7.	Misserwala Kala	19.11	17.46	1.65
8.	Misserwala Khurd	16	15	1
	Total	896.597	844.499	52.098

Source: Field Survey, 2014-15.

These data of field survey as shown in Table 3.3 reveal the diversity in the economy of the region. In contrast to the hilly parts of the district, the economy of the valley region is diversified as depicted in Table 3.3.

Table 3.3: Persons engaged in Agricultural and Allied Activities

Block	*Agriculture*	*Poultry*	*Dairy*	*Fishery*	*Total*
Doiwala	71 19.80%	196 32.60%	67 18.80%	28 23.90%	362 25.26%
Raipur	92 25.60%	124 20.60%	78 21.90%	14 12.00%	308 21.49%
Sahaspur	116 32.30%	93 15.50%	107 30.10%	26 22.20%	342 23.87%
Vikasnagar	80 22.30%	188 31.30%	104 29.20%	49 41.90%	421 29.38%
Total	359	601	356	117	1433

Source: Field Survey, 2014-15.

Rotation of Crops

Restoration of fertility of the field is a prime consideration and the same is achieved by rotation of crops. One year rotation without any fallow is adopted in irrigated field, no matter whether it is the Kharif crop's Mandua or Chaulai and the Rabi crop's wheat or barley. In 'Ukhad' land the ordinary rotations covers a period of two years, which begins with Mandua, sown in May and reaped in October. The land kept fallow until the May of the second year when paddy is sown, followed in October by barley or wheat. For the standard rotation, the lands are divided into two parts; half is sown with Mandua and the other half with paddy. The first half then lies fallow during the 'Rabi' season and the second half is sown with wheat or barley and during the next year the process is reversed. A second well recognized rotation begins with 'Til' (sesamum) followed by 'Masur' (masar) then Kharif pulses and finishing with a period of five months' fallow.

The important fruits grown in the district are the mango, guava, peach, grape, strawberry, pear, lemon and litchi. Dehradun is famous for its litchi. Among vegetables, potato is the most important crop. Potato cultivation in the Mussoorie hills is an old and established industry. Besides, catering to the needs of the town of the district, a considerable portion of the potato is transported to other districts of the State.

Animal Husbandry

As an important food production system this is a complementary proposition to agriculture. The animals like bullocks and buffalo are used as draft for plough; horse, mule, goat, ram as pack or beasts of burden. Cow, buffalo, she goat and ewe as milk and pig, ram, he-goat for meat purpose are very common. Their hair, skin, wool, bones, fat and its manure is utilized for various purposes. Sheep rearing is done commercial by people in the region over centuries, basically for obtaining wool and using as beast of burden for transporting goods from one place to another. For obtaining better wool, milk and other animal products of quality and quantity, government is providing improved pedigree studs and financial assistance to people of the area, to augment animal husbandry.

Trade

The trade in Dehradun valley is well developed and mountainous region of district follows two channels of trade, firstly between the hilly areas and the plains and secondly between the plains and the hills. Timber, charcoal, catechu, fine rice, and vegetables specially potatoes, fruits (litchi, apple, etc.) and wool, etc., are sent to the plains. In return the area receives products of all sorts, cloth, woollen cloths, salt, sugar, foodgrains, fruit, spices, high breed seeds and cattle. All these again are transported to other parts of hills. From where come coarse blankets, rice, ginger, turmeric, red pepper, ringal in various forms, pens, pipe stems, walnuts, apricots, honey, wax, lac, gum, resin, timber and potatoes.

Turmeric, ginger and chillies find a ready market at Kalsi where these are often exchanged with foodgrain in whole or part. Coarse blankets called *lohi* are woven in the hills from the wool of sheep and goats and those in excess are disposed of through the *bania* middleman.

In the upper parts most of the people rear honey bees in their houses. The house walls are furnished with a place in the form of a drawer to accommodate beehive which usually opens outside as a whole which is sufficient for free ingress and egress of bees. They collect and sell the honey along with wax. The dark coloured honey of wild bees is also collected which is consumed by the cattle in place of *gur*.

Walnuts and apricots are grown in most villages and the produce is disposed of either to less favoured neighbours or to the shopkeepers at Chakrata or Kalsi. A few fruit gardens in mountainous region of district do a fair trade with the produce of fruit trees such as apple, plum, pear and the like, which are sold in plains at premium prices.

Division of Labour

As already mentioned the social structure in the district is an exact replica of other part of India which mostly centred around agricultural production systems. The returns of the produces is paid as kind and mostly not as cash for the various kinds of labour operations required at various stages of crop production. As elsewhere in the country Brahmins have been traditionally engaged in worship of gods and the Rajputs are engaged in cultivation. The Scheduled Castes who have been engaged in various kinds of artisan and craft work are now being specialized into sub groups doing mistry in iron and gold smithy, carpentry masonry, pottery, utensil making, etc.

The *hali* (ploughing), *gair* (cattle tender), *bhendal* (sheep tender), *bakarwal* (goat tender), etc., works are being done by persons regardless of caste differentiation.

Land Tenure System

In district of Dehradun each land holder cultivates his own land and there are very few tenants. Cultivators are basically of the

Brahmin and Rajput caste and have the power to sell or otherwise dispose of their lands and have in every way a right and title to their villages. The other category of cultivators is only tenants and has no right to sell or otherwise dispose of their lands. They pay rent to land holder in whose land they cultivate. Situation warrants a common land tenure system applicable to all land holders in the area.

Thrust Areas

1. Emphasis should be on increasing crop productivity by adopting improved agricultural technology. Use of high yielding varieties, use of balanced fertilizers including compost and bio- fertilizer application and adopting timely plant protection, measures are the need of the hour.
2. Production of crops, vegetables, fruits and medicinal plants with improved technology. Medicinal plants have a huge potential as is evident for tremendous success of Patanjali Products.
3. Diversification of agriculture to make small and marginal holdings more remunerative.
4. Increasing domestic animals productivity through improvement in breeds of cattle and buffaloes.
5. Enhance the production of spices, medicinal and aromatic plants.
6. Motivation to local people towards conservation of common property resources and their scientific management.
7. Empowerment of women through their skill development and increase their participation in the decision making.
8. Generate self-employment through agricultural for rural youths.

Chapter 4
Dynamics of Land-Use
Economy from Forest Resources

Forest Resources are in use as common property resources in the district of Dehradun. It describes the rights of persons to enjoy a profit, in common and more generally to describe the land over which their use rights are enjoyed. Each village and township in the district of Dehradun is surrounded by the arable land, cultivated in strips by each householder. Their agriculture is supported by meadows, pasture land and by the waste land which comes under common property (forest) resources. From this forest, each householder is entitled to take wood for repairs and fuel for kitchen. When the fields are under crops and meadows laid up for hey, the waste is used to support the animals and oxen needed for the plow and the cows and sheep which supply manure.

In Dehradun district there are five main categories of returns that can be obtained from forests. Firstly, a pasture is perhaps the most frequent resource, It confers the right to take away the grass and herbage for the cattle. Secondly, a forest gives the right to take fuel for the houses. Thirdly, a forest entitles the people of Dehradun district to take timber for constructing and repairing their houses. Fourthly, forest products and fifthly, the forest land give the rights to extract minor minerals.

The innumerable treasure of forest resources available in the district of Dehradun in the Hmalayan region has a considerable impact on standard of living of its inhabitants which is evident from their food procurement system. Food production system is highly vulnerable to the extremes of climatic conditions. The local people for escaping imminent food crisis have always been keeping supplementary supply from the forest resources, scattered around their habitat in open. Thus various kinds of minerals, spices, condiments, dyes, medicinal plants, fruits, roots, berries, flowers, bark, twigs, gum, resin and a variety of animals, birds and fishes are procured by gathering, hunting and fishing. Apart from the village Panchayat 'Sanjayat' which once belonged to village owners, the rest forest resource once belonged to princely state, is now State property.

In past the Kingship reviewed and settled the tenancy of arable land to the individuals, from time to time and declaring them private property resources (PPRs) owners but kept the rest of the resources intact as reserve to safeguard the natural resources from rapid and ruthless exploitation. From these resources, they allowed the public only to extract timber, fuel, fodder, forage and minor forest products, for domestic use. The same rules of access with minor restrictions are followed today in the district of Dehradun. Commercial exploitation of some of these huge natural resources was extremely controlled and restricted through annual contracting agreements.

The gathering, hunting and fishing are considered to be the techniques of primitive man carried forward by the civilized hill as well as valley people of district Dehradun. Even after learning the process of food production by agriculture, horticulture and domestication of wild animals and birds. The forest and wildlife protection departments of Uttarakhand now regulate hunting and gathering.

Pasturage: The hilly pastures in the district of Dehradun is scattered between village to forest boundaries. The pastorals largely depend on the pastures for pasturing their livestock. The sheep and goat herders pass mostly their time in pastures, who move with their herds year round. In terms of net gains, the pastoral economy stands after agriculture in the region and for some individuals, even more than agriculture as the grass cover is huge which provides nutritious and cost free fodder.

The contribution of forest resources to the life of people in the district of Dehradun and economy should be viewed in the light of the following:

(1) The supply of fodder and grazing space from forest resources help individuals in curtailing expenditure on cattle feed and promotion of health and hygiene of the animals.
(2) Fire wood/fuel wood is in heavy demand in the chilly cold times and used for heating up rooms and cooking purposes daily. The quantum of these woods required for fuel when collected cost free curtails a heavy expenditures.
(3) The forest resources not only supplement food but also medicines, resin and other valuable items cannot be produced by the people commercially.
(4) Forest resources also generate employment opportunities for the poor. As the rich people employ them for bringing fuel and fodder from the forest and they are also employed for taking care of the cattle, sheep and goat in the pastures and meadows.

As per the primary survey it is noticed that the rights in the form of contribution to the economy in Dehradun district are held in the following manner:

(i) ***Open Admittance*:** Free for all use rights of forest resources are neither exclusive nor transferable. These rights are owned by everyone with open access to everyone.
(ii) ***State Property*:** Ownership and management control is held by the State to which use rights, nature of exploitation and access rights are vested.
(iii) ***Community Control*:** Use rights for the resources are held by community of interdependent users. These are not privately owned. These users exclude outsiders while regulating use by the members of local community. The use within the community the rights to resource are of equal access and use.

(iv) ***Private Control***: The right to exclude others from using the forest resource and to regulate the use of the forest resource are vested in an individual or family. Private rights are generally recognized and enforced by the Uttarakhand State. Unlike rights under open access, private property rights usually are exclusive and transferable. The forest resources have certain critical roles to perform in district of Dehradun and that may be summarized under five headings:

(1) *Livelihood Security:* The forest management in the district of Dehradun is traditionally the basic means of ensuring livelihood security to its people. The access rights to vital resource is provided to everyone in the community to avail the opportunity of meeting their basic requirements. Reciprocal obligations for sharing and mutual help appear among the people, viz., fishing, hunting, gathering, etc., as essential traits for group survival.

(2) *Right to Use, Equity and Conflict Resolution:* Forest resources system among the communities of the Dehradun district has an effective mechanism for the equitable use of resource with a minimum of internal conflict. Rules mutually agreed upon and suitable to all members of the group, provide an efficient means of conflicts resolution and in the enforcement of these rules.

(3) *Mode of Production:* The system of Ringal basketry production among the community is done by the member of a particular caste, spread in the different villages of the region. The Ringal work caste members share a common culture, knowledge of the resource and knowledge of resource use rules facilitating production.

The system serves as interface not only between the local society and forest resource but also between the individual and the society at large. Social roles and obligations are often defined in terms of one's participation in work teams. These are an integral part of the culture of the region. Hunting, fishing and fruit gathering, etc., from forest are treated as local ways and means of life.

(4) *Forest and Forest Land Resource Conservation:* The system among the local community are basically conservative in the way resources are utilized. They aim at regular supply and full growth of resource leading to local self-sufficiency. The emphasis is on sustainable exploitation. There are social sanctions against excessive individual collection and accumulation of surplus from a community resource.

(5) *Ecological Sustainability:* The forest resource system deserve credit for many of those resources which have remained productive even after exploitation from generations. The traditional use of resources in the region incorporates rituals such as fishing ritual to help synchronize harvesting with natural cycles. These rituals serve to reinforce social controls in maintaining the productive resource from generations to generations.

CHAPTER 5
Dynamics of Land Markets
Emerging Issues and Nature of Functioning

"Land Market"—Management and Development

Traditional community institutions and statutory Panchayat along with revenue, forest and wild life departments in the district of Dehradun play crucial role, for suitable land resource use, because of their relevance in the development of economic condition of the society. Since these systems form an effective long term and grass root institution framework, It ensures popular participation in development decision making and the issue connected with land market.

Dynamics of land markets in the district of Dehradun has been integrally associated with the political and socioeconomic situation. Having been ignored for longer period in development planning, it is now established through worldwide evidential case studies that the land markets can play a crucial role in economic development. The community opinion on lands earmarked for implementing different development projects, cannot be ignored, as they are the major stakeholders of development in the proposed development plans. The logical approach for development planners would be to deal the land market phenomenon intelligently in tune with socio-economic structure, and also production and resource management issues. However, care should be taken to place land market systems in its proper perspectives. Land markets having developed within specific historical, cultural and ecological contexts, its inherent strength is its suitability for these areas and types. The importance of land market in the area for the sake of development cannot be fully replaced by the traditional economic system but it can be a space for supporting the economy.

While reading previous chapters the scientists, economists and development planners would have realized the importance to develop a proper land market system which is lacking till now in Uttarakhand. There is also a need to overcome the narrow technical perspectives of emerging land-use pattern and the need to become sensitive to the views of others on the subject. The ecological wisdom of the people emphasizes respect, responsibility and protection to land, leading to ecological balance. While implementing a "land market" system ecosystem concerned plan should be ensured, in which the local people should also be a part. An ecosystem view of the people as regards to nature and relations of "land markets" are far richer, experience based, more informative and ultimately more sustainable. In district of Dehradun the "land markets" are live and adapting new challenges wherein sometimes the challenges are disappearing and reappearing due to the want of effective, exploitative and regenerative institutional mechanisms.

Changing and Emerging Land-use Pattern

The word "Dehradun" is a combination of two words "Dehra" which means camp and "Dun" which means valley. The area of district of Dehradun was annexed by the British in 1815AD to the British India and since then it had been one of their favourite places due to its surroundings and climate. This area was the favourite place of Sikh Guru Ram Rai. Britishers developed the region as a major academic and research centre and a base for the Indian Military Academy and the Survey of India. There are also several prestigious boarding schools including the Dun School, India's one of the most exclusive private school. The major factors which contributed to development of Dehradun include the following:

1. Development of the tea industry.
2. The extended operations of the forest department.
3. The growth of Dehradun as a cantonment.
4. As a retreat for well-to-do pensioners.
5. The establishment of Chakrata.
6. The increasing population of Mussoorie.

The Dehradun region has seen a great amount of change in its land-use pattern over the years. During the post-independence period the area has registered an extraordinary growth in its economic aspects. Establishments of large scale industries, e.g., Amitabh Textile Mills Ltd. in 1960, Miniature Bulb Industries of India in 1958, Bengal Immunity in 1969, Dun Valley Combers in 1962, Indian Woollen Textile Mills in 1966, Raj Narain Flour Mills in 1964 and a number of other small scale industries and banks have greatly triggered the growth of the area. Up to the year 1960 there has been no efforts made to monitor and check the messy growth of the region.

For planning, regulation and control of land developmental activities, Dehradun Regulated Area was declared in 1963 under the provision of UP Regulation of Building Operations Act 1958.

As mentioned earlier the district is the most economically developed district of State. However, land market activities are mostly concentrated in the valley region of the district. All urban centres, except the Chakrata cantonment, are located in Dun valley. In Dun valley too, the middle and eastern part is more urbanized than the western part. The Dehradun urban Agglomeration, alone accounts for about more than 70 per cent of total urban population of the district. Except Dehradun, Rishikesh and Mussoorie Urban Agglomerations, all urban centres in the district are of small population base.

A broad and significant land-use change has happened in the last decade and due to this the dynamics of land-use of district of Dehradun has changed especially in the economic factors from original primary sector to secondary and tertiary sectors. Since, land is the main natural resource for life support systems of this area, gradually the land cover changes created the larger problem of regional environmental imbalance in the region.

The changes in land-use and land cover between the years 2000 to 2009 can be seen in Table 5.1.

Table 5.1: Land-Use/Land Cover Classification (2000-2009)

Land Use Class	*2000*		*2009*		*Characteristic Features*
	Area (sq km.)	Area (%)	Area (sq. km.)	Area (%)	
Forest	1071.56	57.98	1032.07	55.85	Forest-Farm Fringe
Agriculture	611.04	33.06	603.59	32.66	Urban Fringe
Barren/Scrub	121.47	6.57	134.86	7.30	Degraded Forest
Built-up Area	30.92	1.67	65.70	3.56	Urbanisation
Waterbodies	13.01	0.70	11.78	0.64	Seasonal Streams
Total	1848.00	100.00	1848.00	100.00	—

Source: Tiwari Kuldeep and Khanduri Kamlesh (2010).

The increasing "land market" pressure is putting tremendous pressure on the land-use and land cover of Dehradun area. The expansion of unorganized land markets has attributed to decline in the area of the land used for agriculture, horticulture and forest activities. The boom in construction sector for industrial, institutional and residential purposes owing to the rapid population growth, has almost quadrupled during five decades from the year 1961 to 2011. Almost 10 per cent of the land-use change is due to the encroachment by slum dwellers and small land developers. The Table 5.2 shows the change in built-up area, over 100 per cent of the total change in area over a decade, which is an alarming trend for the region.

The rapid change in the land-use and attraction of land markets has become significant only after Dehardun became the capital of newly formed Uttarakhand State in 2000. This change in status and the government's changed policy attracted large amount of institutional and commercial activities in the district of Dehardun. This significant change perhaps happened in the last one decade. The Dehardun has developed as main commercial and educational

Table 5.2: Area under Different Land Use/Land Cover Categories Changes (2000-2009)

Land Use Class	*Year 2000*	*Year 2009*	*Changes in Area*
	Area (%)	Area (%)	Area (%)
Forest	57.98	55.85	–3.75
Agriculture	33.06	32.66	–1.3
Barren/Scrub	6.57	7.30	10.09
Built-up	1.67	3.56	112.4
Water	0.70	0.64	–9.5

Source: *Ibid.*

centre for the entire Uttarakhand State and it attracted large amount of population from the hilly tracts as well as from the plains.

The urban growth pattern during five years, i.e., 1987, 1992, 1998, 2003 and 2008 showed double growth rate in the area under study. It can be observed from the year 1998 to 2008, i.e., from 746 ha area to 1463 ha area. A very high growth rate, almost 32 per cent and 49 per cent in 1998-2003 and 2003-2008 respectively, was also noticed during the study (Table 5.3).

Table 5.3: Noticeable Urban Growth Pattern during Five Years

Year	*Area (ha)*	*% change*
Till 1987	561.55	—
1987-1992	690.96	23.05
1992-1998	746.05	7.97
1998-2003	982.64	31.71
2003-2008	1463.41	31.71

Source: Kshama Gupta (2010).

Land Emerging as a Market Commodity

State is the decisive owner of the entire land in the State as per the Doctrine of Eminent Domain. It reserves all the rights to acquire the private as well as divert government land for public purposes. Apart from this, in the open market individuals and agencies also control and work directly and indirectly in the land markets of the district Dehradun. It is going on without any planning. The

Uttarakhand State also retains the rights to mine and quarrying and the other resources on private lands. The local societies of district are also permitted to own freehold land. This type of freehold ownership corresponds to the lack of restrictions to the land owners. Land on lease and rent is also practised by the people who have enough margins and scope of land utilization in their favour.

Ownership rights are further supported by the use rights of the people and such rights are overseen at the local government level. Government can impose physical restrictions through proper planning on the use rights but these restrictions in the use creates negative impact and over administration on the right of the inhabitants. It has been noticed in the district that private rights in land can be exchanged for consideration mostly money by execution of deeds under Registration Act 1908, it can also be exchanged with the other to fulfil some commitment which is acceptable to the both parties through informal agreements.

Access to land for the people is possible only when property rights of lands are held in favour of landowners. Economic investment and any type of production are many affected by the insecure title to land and insecurity of its tenure. When there is a well-defined land right it is easier for "land market" transactions to flourish. In the case of district of Dehradun traditional community land rights systems are working in an efficient way. This is providing effective tenure security with or sometimes without having the formal system of land titles. It is found in study area that insured property rights have increased the investment power of the people and capacity to take risks. Security in the land tenures rights has shown positive support for developing the functioning of the land markets. Now land owners are willing to rent out their land for other than agricultural employment. It also allow poor to earn handsome money rather than depending on earnings out of labour alone.

The prevailing system of land titles does not provide security of land tenures, so full value/potential of the property is not realized. People are afraid to lease out lands, so they are kept fallow or under-utilized. This impacts economic productivity also. So, land markets are not in good shape in the study area.

Functioning of Land Markets

The potential users of the study area can only be given access to the land through a well defined system of market. It is found in the study area that district has different types of land market activities, though there are some similarities. Sometimes Land Markets fails to deliver effectively due to the ignorance of society. In Dehradun district it is noticed that Land Sales Markets and Land Rental Markets both are equally effective in facilitating the beneficiaries. The Land Market of the district has already passed a significant threshold point by now. Land Markets are working quite well as it is seen in the district that Land Sales and Land Rental Markets are equally providing land access to the potential users. Sometimes it is found that legislations are putting pressure on the functioning of the Land Markets. All land transactions are not contributing to the development of Land Markets but cause certain market limitations and loopholes due to some policies of Government.

Land Sales Market: Sales markets of the area are more widespread among Land Market of the district. In these Sales Markets permanent freehold and long-term leases are being exchanged among the parties. Sometimes it is also seen that the sales markets also involve the transfer of rights for a shorter period. For example, if a mortgage agreement is taking place, the second party may do some permanent changes to the land use. This change may be agreed under certain mutual conditions, such as the right to extra construction.

Land Rental Market: Land Rental Market provides use rights to the other party only for a limited period of time. Land Rental markets are found enhancing the productivity and equity by transferring land to more productive users at low cost. These markets are providing opportunities to the landless people to work on land. Indeed the Rental Markets operate in a variety of forms, ranging from highly informal transactions to formalized long term contracts. Facilitation to the less skilled village economy towards the non-farm economy is prevalent in the study area.

The above Sales and Rental Land Markets in the district are basically formal and structured and informal too. In both cases, the financial matters of the land property are depending on the following several factors:

1. Location,
2. Size,
3. Type of project,
4. Type of tenure,
5. Type of land-use, and
6. Current trend in the State.

Land markets influence the income of the people of the district of Dehradun. Regulations of Land Markets in the study area has indicated a negative aspect of hampering the rural growth as evident from Table 5.4. The scopes of access for landless is very much limited and even the land owners are not motivated to rent out their lands, reducing the productivity and investment on land for agricultural purposes. The percentage of share and contribution is shown in the following Table 5.4.

Table 5.4: The Percentage Share and Contribution

Sl. No.	*Contribution*	*Type of Contribution*	*No. of Respondents*	*Percent-age*
1.	Agriculture	Productivity and efficiency	13	10%
2.	Industry	Investment and development	52	40%
3.	Individuals	Borrow power and employment	26	20%
4.	State Government	Revenue	39	30%
	Total Respondents		130	100%

Source: Field Survey, 2014-15.

In urban areas of the district, Land Markets are creating new opportunities for the migration of the local people, mainly for finding labour. Land markets facilitate the people from rural communities to migrate to the urban business centres of Vikasnagar,

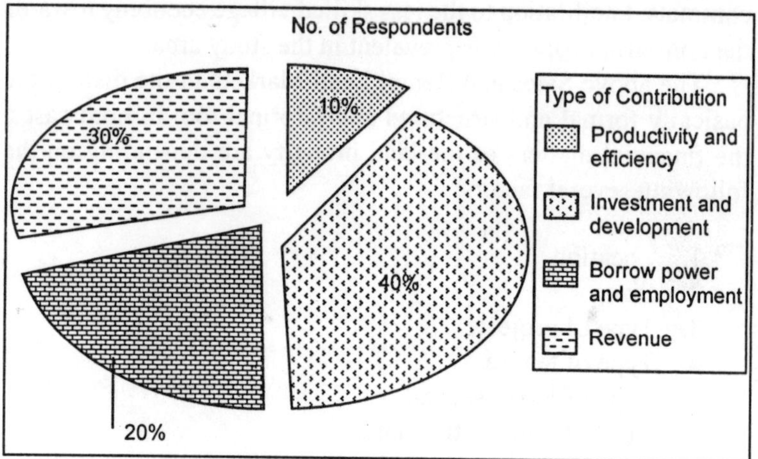

Fig. 5.1: Number of Respondents and their Type of Contributions

Sahaspur, Dehradun and Rishikesh and Doiwala. Land Markets are supporting the livelihood of people and providing them money for running small business activities. Land Markets are also influencing the various other socio-economic characteristics in the area as given in Table 5.5.

Table 5.5: Land Markets influencing various Socio-economic Characteristics

Sl. No.	*Socio-Economic Impacts of Land Markets*	*Significance given by the Respondents (No. of Respondents)*	*Significance given by the Respondents (In %)*
1.	Attitude	11	8.24
2.	Standard of living	30	23.24
3.	Material culture	19	14.62
4.	Government policies	4	3.00
5.	Financial Support Services	14	10.07
6.	Infrastructure	27	20.80
7.	Shift in Political Decision Making	13	10.00
8.	Land-Use	12	10.01
	Total Respondents	130	100%

Source: Field Survey, 2014-15.

In Dun Valley region great degree of unpredictability is seen in the overall nature of land markets in tune with the State economy and National economy. The land owners and potential users involved in the land market of Dehradun region are facing risks in

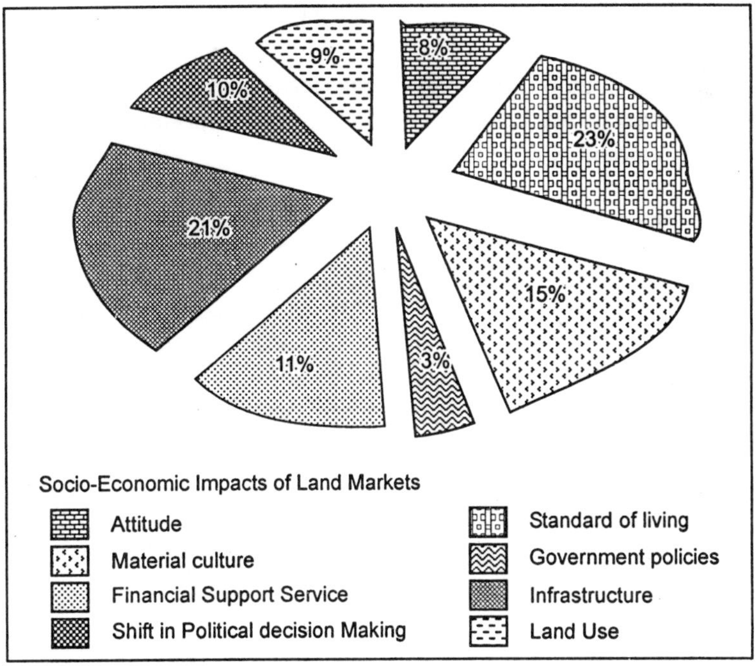

Fig. 5.2: Significance given by the Respondents (No. of Respondents)

terms of property values that may go sometimes up and sometimes go down as well due to unpredictability. Indirectly or directly individuals invest money in the operations of the land market which comes from other economic activities mainly the surplus funds.

Importance of Land Markets

Up to 1970, it was a common thinking among the planners that safe and sound land tenure for the communities is important and it is sufficient as the source of income to generate an environment for economic development. Now there is a shift in this view. Secure tenure alone is not adequate to cope up with the requirements of people now a days. They need new dimensions of fast earning and hence land markets came into action in district of Dehradun as per the demand of time. In the study area good quality of land related information is available and readily accessible. Though complexity is increased in the processes, the need for technology and support has been observed as critical. In Table 5.6 reasons attributed to high magnitude of land market transactions are given rank-wise. These reasons are compiled on the basis of importance given by the respondents of the study area.

Table 5.6: Reasons attributed to High Magnitude of Land Market Transactions

Sl. No.	*Reasons*	*Importance/ Ranking (No. of Respondents)*	*Importance/ Ranking (In %)*
1.	Land Availability for all Development Requirements	36	27.89%
2.	Economic Benefits	22	16.91%
3.	Social Benefits	1	00.17%
4.	Global Competitiveness	2	1.65%
5.	Technological advancement of the area	14	11.00%
6.	Trade development	5	3.34%
7.	Breaking down Barriers of Economic Development	18	14.56%
8.	Breaking down Barriers of Infrastructure development	10	07.54%
9.	Providing National and International Linkage to the Peasants of Study Area	5	3.55%
10.	Participation of Local People in the National Economy	11	08.27%
11.	Equal allocation of Rights to the Rural People	6	5.12%
	Total Respondents	130	100%

Source: Field Survey, 2014-15.

The economic development in the study area basically depends on degree of availability of land. For any office there is a need for land for making a workspace. For industries, Factories, Roads, Railways, Housing Societies, Markets and Retail Centres needs a sufficient quantity of the land for its establishment. Therefore, in the study area it is found that land market is a key factor in developing a new source of livelihood. Every person in the district observes land is a valuable commodity. The population of Dehradun district is expanding day by day but since land is a limited commodity, there are problems in fulfilling the increasing demands of land. The problem of sufficient land is found as a very delicate situation particularly in Dehradun Valley region.

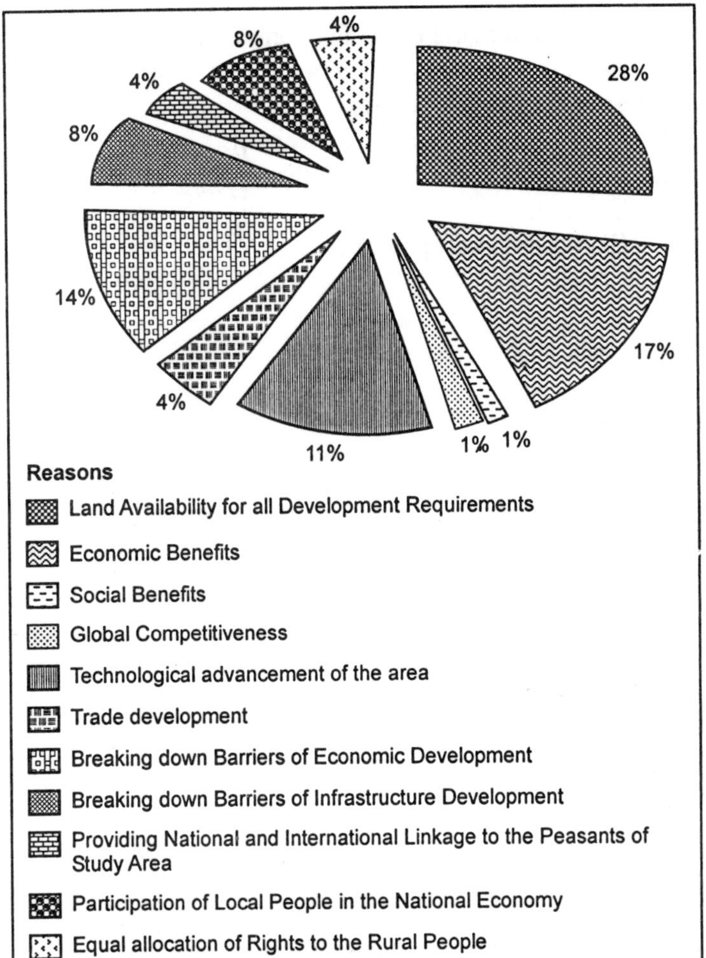

Fig. 5.3: Importance/Ranking (No. of Respondents)

Despite the fact that land markets are functioning in unorganized manner in the district of Dehradun, these are not much destructive for the flora and fauna of the region so far. The development in ICT need to be harnessed, for the seamless transaction of land in a transparent and easy environment. The following immediate requirements were observed in the study area:

1. Promotion of the Local Economy.
2. Breaking down the barriers of cross community income groups and class lending.
3. More opportunities and safe credit.
4. Providing economy link to the rural people of district of Dehradun.
5. Exposure to best practices.

CHAPTER 6
Status of Land Markets, Productivity and Magnitude

Factors affecting Participation

Land Markets of the study area are supporting the economy of the people through inherent mechanisms and these Land Markets help people in creating assets, in a suitable manner. Being given protected land privileges in the study area, land markets can amplify the motivation to investment and loan for assured guarantee of income generation and production. In the 9 villages of Sahaspur block of the district of Dehradun improved economy and prosperity was found where the land markets are prevalent (Table 6.1).

Table 6.1: Magnitude of Traditional and "Land Market" Economic Activities

Mode of Occupation	*No. of Persons Involved*	*Area Engaged (Acres)*	*Average Income (in Rs. per Month)*
Agriculture	55	6.00	15000
Dairy	12	0.50	25000
Fishery	14	1.21	17000
Poultry	16	0.89	30000
Real Estate	8	21.35	60000
Factory	2	130.87	50000
Other Business	50	7.841	20000
Service	700	0	30000

Source: Field Survey, 2014-15.

In comparison to other traditional economic activities the earning from Land Market activities was observed significant as given in Table 6.1. In Real Estate and Factory activities only 10 persons were seen engaged, but both activities generate more than Rs. 1.00 lakh per month. The average share of income per head is more than Rs. 10,000. The levels of benefit and the impact of Land Markets are seen very much positive and noticeable in the study area. Land sales among small holders was rare and in general, persons generally recognized land rights through the Land Market while land rights that are acquired through customary practices of gift or inheritance falls under the community or customary tenure. In the study area the Land Markets have been found little biased towards economically sound people and which is evidently sidelines enough number of poor and landless people.

Assessment of Land Transactions

The value of land is found increased manifold under the functioning of Land Market in the study area. The purchase rate of the land is evaluated on the basis of the purpose of economic activities. Therefore, the price of land increased in favour of land owners. State government have fixed circle rates for each type of land depending of its location and distance from the highways. The Land Owners were getting more amount of money from the prescribed circles rates fixed by the government most of the times. The open market prices are always found different from the government rates. This scenario of land sale and purchase in the

Table 6.2: Market Rates of Land for Different Purposes

Sl. No.	*Road/Area (Mussoorie)*	*Year 2012-2013 (Sq. metre)*			
		Residential Rates (Rs.)	*Commercial Rates (Rs.)*	*Office Rates (Rs.)*	*Hotel Rates (Rs.)*
1.	Library point to Solaire Plaza	11000	14300	130	90
2.	Company Garden to Hathipaon	2500	2800	110	90
3.	Dick Road	2500	2800	110	90
4.	Library point to Indra Bhawan	6600	10500	130	90
5.	Charlivilli to Tibetan Homes	2200	2800	110	90
6.	South Road	1750	2400	110	80

Source: Registration Department, Uttarakhand.

study area gives more scope to better assessment of land price. The market rates given in Table 6.2 shows the market rates of land for different purposes.

During the land sale and purchase, the rates of land taken much higher than the fixed rates by the State government as shown in the Table 6.2. The decision by State government regarding market value of land always determines the indemnity of land transactions and it has significant role in land markets in the district of Dehradun. The market value of land is also assessed when compensation needs to be paid for land that is compulsorily acquired by the State Government. The land market efficiency depends on how many people exchange their lands into "land market" related economy and for other than agricultural purposes. In the study area at least 40 per cent capital generation is seen possible through Land Markets.

In the district of Dehradun, Land Market is seen operating in the form of land tenure transaction most of the time. Since the agriculture no longer feed the landless labour and marginal farmers, land markets offer, skilled labour and credit facility to the inhabitants of the study area. It is fact that behind the land sale by the small and marginal farmers, there is pressure exerted by increased demand on food, education, shelter, etc.

In the land market prone areas, i.e., Doiwala, Raipur, Sahaspur and Vikasnagar blocks of district of Dehradun, it is observed that upper castes and big land owners have ownership and control on the most of the land but they themselves are not cultivating the land. It does not mean that they keep their lands fallow. These land owners sometimes lease out and sometimes on contract agreement cultivate their lands. On the other hand, there are small, marginal and landless people who have the skill to do cultivation but due to lack of resources and small area of land ownership they cannot utilize their skill for generating sufficient income through agricultural activities. There is another big landowner group found in the study area, who with the help of landless labours actively do agricultural activities. The marginal and small landowners are dependent on their family members for meeting labour requirements in the agricultural fields.

The total number of land transactions in 16 studied sample villages were 2,755 and the total land areas under land sale and purchase was 31.05 acres. In Table 6.3 it can be seen that maximum number of land sale and purchase cases observed in village Jakhan. This is because this village is situated on Dehradun-Mussoorie-Yamunotri Highway. Here scope for real estate and business is comparatively high. The same case is seen in Brahman Gaon village which ranks second in the number of land transactions. Few villages as shown in Table 6.3 have witnessed no land sale and purchase cases. It is due to their issues in the location and distance from the highway and cities of the district. Therefore, distance and location of land also has a role in the functioning of the land markets.

Table 6.3: Ratio of Land Sale and Purchase (From Year 2010 to 2014)

Sl. No.	*Name of Village*	*Cases of Land Sale*	*Area (acres)*	*Cases of Land Purchase*	*Area (acres)*
1.	Jakhan	1800	10	1800	10
2.	Kishanpur	500	2	500	2
3.	Pratitpur Santour	6	0.5	6	0.5
4.	Dhak Patti	300	2	300	2
5.	Rajpur Maphi	20	1.5	20	1.5
6.	Veer Girwali	7	0.8	7	0.8
7.	Hathdiwala	0	0	0	0
8.	Makdeti	20	1	20	1
9.	Thani Gaon	15	0.75	15	0.75
10.	Dungal Gaon	5	0.4	5	0.4
11.	Kirsali	0	0	0	0
12.	Danioka Danda	10	0.8	10	0.8
13.	Kuthal Gaon	7	0.5	7	0.5
14.	Bagral Gaon	50	4	50	4
15.	Makkawala	0	0	0	0
16.	Brahman Gaon	15	6.8	15	6.8
Total		2755	31.05	2755	31.05

Source: Field Survey, 2014-15.

Table 6.3 also indicates that land sale cases and land purchase cases both operate at equal intensity and frequency. The difference is that land sale players are from the village and land purchase players are from outside from the village. It is also a fact that big land owners are purchasers of land and small and marginal land owners are the major sellers of land in the land markets.

The flow of capital is purely internal in the Land Markets of the study area. In Table 6.3, sixteen sample villages of block Sahaspur, district Dehradun were surveyed and observed that the total 2,755 cases of land sale and land purchase were found, involving 31.05 acres of land area. The benefit of "land market" activities goes to a maximum number of native land owners and purchasers.

Status of Land Administration

Land Administration is a basic element which is decisive for recording and distribution of information about the ownership, circle rate of land and defining the land-use, etc. It plays important role during implementation of land administration policies. Time and money is associated with the registration of land so there is an urgent need to improve the land administration system. Land titling can be an important component as developers may face serious challenges during transactions due to difficulty in finding accurate and sufficient land information.

There should be an attempt to reduce the unnecessary delay in Land Mutations as it put direct pressure on the Land Market transactions. Sometimes it is seen that land sale parties may change their agreement and it is a demotivating factor for other transactions under land markets. Because of undue delay in obtaining the mutations of their purchased land based on its registry, there is need to develop a system wherein legitimate landowner get their mutation done for their land within a week.

Being the *bona fide* owner of the land, once transfer of registry or mutation is done the new owner can start his work and invest his capital confidently on the land. It is observed during field survey that there are 100 to 250 such cases pending in the offices of the land registrar in the district. The situation of Land Mutations and its frequency can be seen in the Table 6.4. It also indicates the pace of "land market" activities in 7 rural villages of Raipur block.

Table 6.4: Raipur Block: Frequency of Mutations

Sl. No.	*Name of the Village*	*Year*	*No. of Mutations*	*Method of Transfer*
1.	Chalang	2011 to 2013	40	Sale deed
2.	Khurawa	2011 to 2013	30	Sale deed
3.	Nagal Harnala	2011 to 2013	10	Sale deed
4.	Marotha	2011 to 2013	12	Sale deed
5.	Vadawali	2011 to 2013	10	Sale deed
6.	Kulhan Karanpur	2011 to 2013	8	Sale deed
7.	Kulhan Mansingh	2011 to 2013	8	Sale deed
Total 2011 to 2013			118	Sale deed

Source: Field Survey, 2014-15.

Land Markets also give space to the mediators to intervene and work in the following ways:

1. Mediate in selection of the land as per the requirement of the party.
2. Mediate in administrative help in conversion of land-use in case the land under sale or lease/rent may set off for non-agricultural business purposes.
3. Mediate in Registration Office for early registry and land mutation in time.
4. Provides space for registered architectures for design land for better economic activity.
5. Mediate in providing loan from banks and financial institutional help to approve the maps, etc., from the Mussoorie-Dehradun Development Authority (MDDA).

The activities going on for mediation in Land Markets are illegal though the mediators are on demand in practice. These activities of mediators are only surviving as they are facilitating the market players coming from outside the district (Table 6.5).

Table 6.5: Mediation in Land Transactions

	Doiwala	*Raipur*	*Sahaspur*	*Vikasnagar*	*Total*
Mediated	4 30.80%	1 7.70%	4 30.80%	4 30.80%	13 100.00%
Not Mediated	160 10.20%	500 31.70%	417 26.50%	499 31.70%	1576 100.00%

Source: Field Survey, 2014-15.

The Gram Pradhan should be the first tier of decision-making authority in case of land-use change in the "land market" transactions. As a Gram Panchayat includes multiple revenue villages, it should look after the motive of agricultural land sale in land markets. It is also important to keep check in the land transactions in other words allow only individual land claims. This is also their responsibility to protect the rights regarding right to administer, look after and preserve village agriculture and forests. Sometimes the Land Markets forces governments in the enthusiastic development of "land market" activities deny or ignore these community rights emotionally. It is observed that in the study area the community rights of land continue to be neglected, and claims for such rights are being actively discouraged. Preference is given only to the individual rights.

Table 6.6: District Dehradun: Role of Gram Pradhan/Gram Sabha

	Doiwala	*Raipur*	*Sahaspur*	*Vikasnagar*	*Total*
Yes	2 50.0%	1 25.0%	1 25.0%	0 0	4 100.0%
No	486 23.2%	514 24.5%	582 27.8%	514 24.5%	2096 100.0%

Source: Field Survey, 2014-15.

To observe the role of Gram Pradhan in the present study a total of 2,096 villagers of four blocks were surveyed. Out of 4 blocks only Vikasnagar block has shown negative response in the case of any role at Gram Panchayat level. Remaining 3 blocks have shown positive response towards the constructive role of the Gram Sabha in "land market" activities (Table 6.6).

Government Revenue from Land Markets

It is a fact that as more and more Land Market transaction will take place government will positively be benefited with the income through revenue collection. Land prices can be clearly decided and negotiated among the stakeholders under the course of action of Land Market activity. In the study area tax on land is directly connected to the Government revenue and indirectly infrastructure development for the people. Uttarakhand government collects direct taxes and stamp duties during land transactions. Stamp duties are imposed under Indian Stamp Act 1899. Thus direct tax on the worth of instruments used in various "land market" business and property transactions is one of major source of income for the government. The stamp duty is two types:

1. ***Judicial Stamp Duties*:** This fee collected for legal proceedings of land transactions.
2. ***Non-judicial Stamp Duties*:** This is imposed for one-time on transfer of immovable land property.

State government provides services for recording the contracts and deed. When any transaction happens during the functioning of the Land Market, State government collect an amount as registration fee for registering the land property. Irony is this, that this registration is not a guarantee of land titling. In the surveyed district of Dehradun, it was seen that registration fees and non-judicial stamp duties account for nearly 90 per cent of revenue collected by the Department. In Table 6.7 (A) and Table 6.7 (B); and Figure 6.1 and Figure 6.2 of revenue collection for the year 2006, 2007, 2009, 2010, 2011, 2012 and 2013 are given. It is evident from these tables that revenue collection during the Land Market activities is an important source of income for the government.

Increase in registration rates from time to time by the Uttarakhand government has increased the revenue collection and income. State government is directly and indirectly benefited by collection of the revenue through transactions in Land Markets. In the study area, i.e., district Dehradun it is investigated that though

Table 6.7 (A): Revenue from Registration of Land (Financial Year 2006-2007)

Sl. No	*Registration with possession*		*Registration without possession*		*Patta*		*Vasiyat*	
	No.	*Income*	*No.*	*Income*	*No.*	*Income*	*No.*	*Income*
Dehradun-I	17	15627100	166	23965370	81	6153050	457	50270
Dehradun-II	26	4231950	154	15089070	94	15173755	574	76110
Vikasnagar	3	122320	231	13084380	83	12955330	202	22220
Rishikesh	1	29510	170	5461375	48	3174110	189	20790
Mussoorie	—	—	—	—	10	974280	5	550
Total	47	20010880	721	57600195	316	38440525	1427	169940

Table 6.7 (B): Revenue from Registration of Land (Financial Year 2006-2007)

Mukhtarnama		*File*		*Others*		*Grand Total*	
No.	*Income*	*No.*	*Income*	*No.*	*Income*	*No.*	*Income*
1746	243890	2	10000	591	3828735	15528	815288029
2172	342980	61	2500	515	23633060	13969	710393320
551	65160	545	80350	366	2322992	13728	414258012
604	77060	274	241900	76	452655	8378	308977130
20	4550	—	—	15	27538	266	33031238
5093	733640	882	334750	1563	30264980	51869	2281947729

Source: Registration Department Dehradun.

there are high rates prevailing in the area, the buying and selling of land has shown an increase in the number of land registrations.

Land Markets stop slow valuation of the land which is under transaction. The registration fees are a certain percentage of land rates, it lowers the burden of land owners and it also minimises the under valuation of property. As it is evident from Tables 6.7 (A) and 6.7 (B), 6.8 and 6.9 the registration department's earning are the main source of income for the government. During the year 2006-2007 as shown in the Table 6.7 (A) and 6.7 (B), the total earning in district of Dehradun is Rs. 228.15 crore from registry for various purposes through "land market" activity. The earning from registration and stamp duty is Rs. 61.81 crore which is 104.59 per cent of the target fixed by the Sub-Registrar Dehradun-I for the financial year 2012-2013 (Table 6.9).

Table 6.8: Office of the Sub-Registrar, Dehradun-II: Details of Income Sources and Number of Lekhpatra (April 5 to 5 March, 2010)

Time	*Financial Year 2009-2010*				*Financial Year 2010-2011*				
	No. of agreement	*Income from Stamp Duty*	*Income from Registration Fee*	*Total*	*No. of agreement*	*Income from Stamp Duty*	*Income from Registration Fee*	*Total*	*E.P.F.*
Upto April	10144	475470955	38135380	513606335	12641	410403830	60969225	471373055	2464460
Upto March 2009	1216	49663725	5145360	54809085	1168	37540910	6996800	44537710	288070
Total	11360	525134680	43280740	568415420	13809	447944740	67966025	515910765	2752530

Table 6.9: Office of the Sub-Registrar, Dehradun-I: Details of Income Sources and Number of Lekhpatra (Upto March 2013)

Time	*Financial Year 2011-2012*				*Financial Year 2012-2013*				
	No. of agreement	*Income from Stamp Duty*	*Income from Registration Fee*	*Total*	*No. of agreement*	*Income from Stamp Duty*	*Income from Registration Fee*	*Total*	*E.P.F.*
Upto April	11848	365975735	78855975	444831710	11265	473592980	87944380	561537360	2355510
Upto March 2009	1532	50638110	11663215	62301325	1104	4748995	9121035	56611030	248020
Total	13380	416613845	30519190	507133035	12369	521082975	97065415	618148390	2603530

State government's collection of stamp duties and registration fees has a positive effect on other taxes imposed by the government as well. It is also seen that transaction of land properties are deferred, from one to another and though formally registered deeds might be undervalued. Both sale and purchase intentionally do this under valuation they pay a part of the deal in cash in such transactions. This type of money comes under undeclared money in "land market" transactions.

Government has fixed circle rates for land property whereas "land market" indirectly decides the commercial price of land. When private land is being acquired by the State government for development purpose, it is required to calculate the rate of that land property. In that case "land markets" made a difference in favour of the local people. Sometimes it is observed that land prices are inaccurately decided by avoiding the Land Market prices. This gives scope to dispute and community dissatisfaction.

Due to the prevalent functioning of the Uttarakhand Land Administration system a direction has been seen in the functioning of Dehradun "land markets". This can be understood by the following illustration:

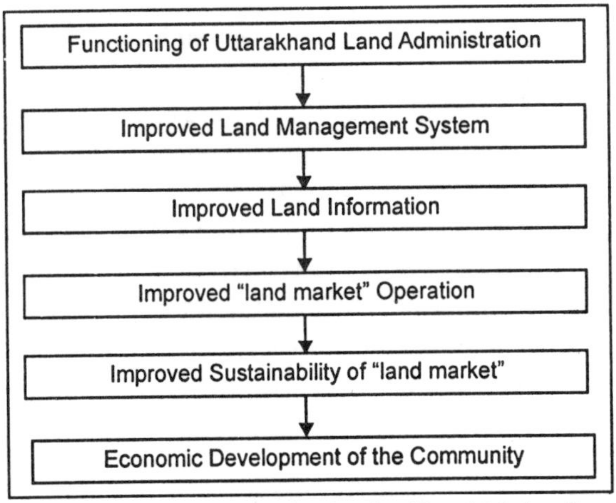

Fig. 6.1: Functioning of Dehradun "Land Markets"

Land Markets are promoting better entrepreneurships and economic development in the locality. It has reduced the responsibility and pressure of the State government substantially. Land-use is purely for the benefit of the local people and community at par with their demands. Land Markets are creating better conditions in the district and State by promoting social and economic factors. Better civil amenities are also created due to the development of land markets in the district of Dehradun.

Land Markets are promoting the community earnings and living standard of the people of the study area due to the favourable conditions as noticed in the study area. These conditions are more suitable in the Dun Valley region in comparison to the hill areas of the district. Some of these are:

1. Awareness of the Land Rights.
2. Credit facility.
3. Affordable circle rates and stamp duty.
4. Educated population.
5. Understanding of "land markets".
6. Cultural and Social recognition.
7. Transparent system of defining the land-use of the area.
8. Political Support.

In the study area illegal development of informal Land Market transactions is observed due to a plethora of factors. The main reasons behind informal Land Markets are basically motivated by poor land records management, opaque system of land-use changes, corruption and lack of political will, etc. Migration from rural to urban and from hills to valley region of the Dun gave entry to the illegal transactions in the Land Markets. Corruption in the system is responsible for the growth of illegal colonization which is located in the forest lands, river banks, waste lands and other government lands. Vote Bank politics is hampering the whole functioning of Land Markets and benefits to a limited number of people.

Uncertainties of the Land Markets

There are fears which retain differences with regard to systems of land tenure while allowing it for the open exchange of land-use under Land Markets. Regarding land property there is a some degree of fear in the study area. Opening Land Market to the outside people gives rise to distress in the land owner community. They feel that this will adversely affect their culture and that they will not be able to grow crops sustainably for their future generations. However, the tendency towards Land Market related economic activities and industrialization in line with international developments rather than compressed in a State framework. Management of land for economic development is necessary to invite investments from outside. The local people also fear that their role in the "land market" is very minimal. Once they sell their land they might be out of the stream permanently and will not be treated as partners in future.

Government Intervention

The State government needs to intervene for facilitating the proper use of the land in a way that ensures sustainable development of "land market" in the district. It will insure and sustain the optimum socio-economic benefit, for the present generation and make it available in a sustainable way for the coming generations.

There are few areas found under which the State government can work in the dynamics of the land market in the area:

1. Removal of market limitation and malfunctions of land markets to improve their effective function.
2. Removal of mediation.
3. Inclusion of disadvantaged groups and landless people in the whole process of "land market" so that sharing of resources can be improved in all social groups.

Role of Women and Disadvantageous Sections

Under Constitution all citizens have equal rights. The role of women in the Land Markets is minimal as traditional customary laws and practices have a discriminatory view towards them. This situation of women in the "land market" functioning is little bit better in the hill regions of Mussoorie and Jaunsar-Bawar. They took part in decision making in property issues. In Dun Valley region they are almost out from the "land market" system. Irony is that despite women of Dun Valley region are more educated. They are not given due role in the Land Markets. The operations and growth of Land Markets are unbalanced without the participation of women. While it is observed that though women face restrictions in the Land Markets activities they still contribute and play their role indirectly.

In Land Markets there is current demand to ensure participation of women and disadvantageous groups on equal basis. At community level society has to see how the customary laws can force the women, weaker sections and minorities on the ownership issues. This issue was neglected at the political level and community level since very long time. In the study area there is a common thinking that Land Markets entry to rich and landowners.

Role of Professionals

The Land Markets of district of Dehradun are functioning with the help of skilled and professional people of the Dun valley region. In district of Dehradun out of total 466 samples, Block-wise list of professionals and technical graduates involved in land market activities are given below. It shows that percentage of potential professionals is at significant level in comparison with other educated people.

Table 6.10: Percentage of Educated/Professionals in District Dehradun

Block	*Primary*	*High School*	*Inter-mediate*	*BA/MA/ Professionals*	*Technical Graduate/ Professionals*	*Total*
Doiwala	84 17.6%	194 24.9%	92 30.7%	91 42.7%	5 45.5%	466 26.0%
Raipur	119 24.9%	193 24.8%	79 26.3%	47 22.1%	1 9.1%	439 25.0%
Sahaspur	124 25.9%	244 31.3%	86 28.7%	46 21.6%	2 18.2%	502 28.0%
Vikasnagar	151 31.6%	148 19.0%	43 14.3%	29 13.6%	3 27.3%	374 21.0%
Total	478 100%	779 100%	300 100%	213 100%	11 100%	1781 100%

These professionals are involved in planning and operations of Land Sales Markets and Land Rental Markets. The role at political level and planning level is significant as they contribute in policy making and also give an ear to the "land market" related problems. Integrity, competence and competitiveness of them significantly help the development of Land Markets.

CHAPTER 7
Concluding Remarks

Land is one of the most durable resources. It is considered as a guarantee and provides safety and sustainability to generations. Land ownership brings a sense of deep rooted identity and a source of political importance with sustainable economic returns. In district of Dehradun of the State of Uttarakhand which was the study area, certain peculiar characteristics of land are present, such as unique physical features of each portion of land, absence of mobility, and prevalent emotional attachment of land owners to their land which contributes negatively and positively in the functioning of land market. Development patterns in Uttarakhand have become increasingly important as State became the centre for investment and the economic transformation after its creation in the year 2000. There are peculiar characteristics seen with the functioning of land markets, which have been exacerbated by Uttarakhand State government by its interventions in achieving orderly growth of industries and urban economic areas. These policies and interventions, together with the viewing of land as a source of revenue, have resulted in rapid development of land markets.

In the district of Dehradun land markets are playing important role to reduce social poverty and inequality. Land Markets help find land to the industries, companies, institutions and real estate in such a huge way that these are changing the land-use pattern of the district. It is observed that the markets of land sales are enhancing the economic status of the rich people now comes from other than agricultural sectors and very less people from land owner communities are getting benefit of it and rates involved in it. Most of the times there is agreement for permanent transfer of land ownership and rights. In this way one is selling party who is handing over his permanent centuries old rights, at one time payment of capital, to the other party. After that first party has no role in the land market scenario. Land rental or lease markets of the area are bit different as these are associated permanently with the land owner. These markets are pro-poor and through this market local people are feeling always part of the land market system. Whereas land sales markets need big capital investment where poor land owner have no other option but to sell his land and go out of the system. Land contracts and agreements are giving easy right of entry to the local poor people in the land market either by contracts or agreements. Policy framework of the State government of Uttarakhand is also encouraging land lease markets and it reflects in better functioning of rental markets in the study area.

It is observed in the study area that when local people face insecurity in traditional agricultural system of food production system they tend to move towards new opportunities of more productive occupations. Functioning of land markets are giving less costly ways to realize land ownership by them. Agricultural labour now has option to move towards the more earning occupations especially the income from Land Markets. Land Markets are proved good for all economic groups and for the economic development of land owners, agricultural labour and landless people of the area. As Income generation in other than agricultural occupations needs skill development. The people in the age-group of 18 to 45 years are more fitting land market associated occupations. And the people of the higher age-group, if not able to participate as skilled workers and as professional, they may participate in various associated functions in the land market systems. It is evident from the study that there are no restrictions for different economic groups in the operations of land markets.

It is interesting to note that in the study area land markets have ability to utilize the transfer of land from less productive area of agriculture to efficient enterprises. At some level there are imperfections in the credit and insurance facility in the land markets of the area. It is also noticed that investment made by professionals have made land markets functioning and improved their operations in support of inhabitants of the study area. On the other side despite so many efforts made the traditional government interventions are still not doing much for the landless and land owner communities.

Since 1991, the economic reforms have opened up the district's massive Land Market growth potential and invited powerful entrepreneurial market forces. The study finds that with the economic reform in the district of Dehradun, the land markets helped to reduce landlessness. The marginal and medium sized land parcels started becoming part of land market economic activities. The Land Market especially sale and rental market is under functional due to variables in the study area, including small land size, high level of education, enough agricultural diversification, cash crops and animal husbandry related incomes. Due to this reason land leasing is also becoming scarce due to unhealthy agreements. Those Land Market outputs, inputs, rent activities and employment is showing diversity of economy, the primary dependence on agriculture is still continuing. Land Market show favourable and positive impact on land, labour and employment.

The Land Markets of the study are reallocating the marginal and economically viable lands for benefits of the small farmers. The legislative restrictions on land are maintained to protect the rights of land ownership, with also sometimes has a role in lowering down the high transaction costs in land sale and purchase that fall excessively on the rural poor. It is because the reason that establishment of Land Markets has not been given much importance and almost not any reform taken place systematically. External investors sometimes worry about the quality and quantity of land dealers, rights and liabilities throughout the negotiations. Sometimes the local land deals dragged up to the courts and land clash continued. In this way it affects the real land value and investment in Land Markets that turn out to be low in the study area.

Proper land titling system and security of tenure is required to increase in land prices and it will also help to bring better profits in Land Markets. State government of Uttarakhand would also be benefited in the form of stamp duty, registration fee and other taxes, etc. Easy and investor friendly land transactions will go ahead to develop the economy of Land Markets in the form of industrialization, urbanization and real estate business.

Assured safekeeping of Land Rights is of utmost importance as the Land Markets functioning in the district of Dehradun, straight robust system to protect land rights of the people. Firstly, there is need for the formal recognition by issuance of Land Titles by the Uttarakhand State; and secondly, facilitation by the State government for smooth transaction of deals in rights, through the process of land registration. Though it is compulsory to register the sale and purchase of land, the Registration Act 1908, does not ask the registration authority to confirm the past history of the land or ownership. This situation restricts the buyer to safely invest in land and it becomes his own responsibility to verify the land ownership of the seller. His deal in land property is always remains questionable and that can be challenged on numerous fronts. So, the investor has to factor in such risks before proceeding with any investment.

The Land Markets should be provided with access to credit facility. Guidelines and institutions are required to ensure credit mobilization and reasonable agreement for land transaction. The study suggests that large source of revenue has been generated in the Land Market activity that can be invested by the district authorities to develop adequate infrastructure to provide credit access to the local people. For landless people the potential benefits of the land market is limited largely on to the labour market opportunities. It is not due to the lack of available land instead the lack of security and transparency of land rights that affects the credit access of people for the Land Market activities.

There is need to move towards low land transaction costs for all users. Further, there should be transparency, openness, ease of access to all in the land market and there should be protection of interest of marginalized social groups. Appropriate legislative measure and strengthening of land administration is required to harness full potential of the land markets. Awareness among various stakeholders especially policy makers about the cascading effect of land markets on all sectors of economy will certainly give a boost to this endeavour. In fact an integrated strategy needs to be put in place to efficiently harness the land markets for increasing revenue of the government and ensuring socio-economic development particularly to vulnerable sections along with environmental sustainability. Increased government revenue from efficient land markets can be further used for various developmental activities aimed at ensuring inclusive growth.

It is widely expected that India is going to be leading economy by 2050 and if at all it takes longer to reach the top position in the economic sector, one of the key impending factor would be the training real estate sector, which is marred by market inhibitions. These market inhibitions are mainly posed by the redundant laws and systems which restrict a seamless and error free land transactions in a transparent environment without much hassles. With the help of it, Land Markets play a vital role in pushing the government to adopt new strategies towards making the land business easy.

Appendices

उत्तर प्रदेश स्टाम्प (सम्पत्ति मूल्यांकन) नियमावली 1997 के द्वारा प्रदत्त अधिकारों का प्रयोग करते हुए जनपद देहरादून के तहसील देहरादून के लिए द्विवार्षिक न्यूनतम औसत बाजारी दरों का निर्धारण विवरण पत्र के अनुसार किया जा रहा है। यह दरें दिनांक 1 नवम्बर 2003 यथा संशोधित दिनांक 2 दिसम्बर 2003 से दिनांक 31 अक्टूबर 2005 ई. तक प्रभावी रहेंगी।

नगर निगम में पूर्ण रूप से सम्मिलित ग्राम की दरें

क्र. सं.	क्षेत्र/राजस्व ग्रामों के नाम	सिंचित दर प्रति हैक्टेयर (लाख रु. में)	असिंचित दर प्रति हैक्टेयर (लाख रु. में)	1000 वर्ग मीटर तक की भूमि के अन्तरण के लिए निर्धारित दरें (प्रति वर्ग मीटर)				व्यावसायिक दर प्रति वर्ग मीटर	दुकान/ रेस्टोरेंट/ कार्या. की प्रति वर्ग मीटर दर	अन्य व्यवसायिक प्रतिष्ठान की दर (प्रति वर्ग मीटर)		
										गोदाम	होटल/ गेस्ट हाउस/ बारातघर/ सिनेमा हॉल/ पैट्रोल पम्प	अस्पताल/ धर्मशाला/ मनोरंजन स्थल जो कि उ.प्र. वाणि. अधि. के अन्तर्गत है।
		निर्धारित दर	निर्धारित दर	500 वर्ग मीटर तक सड़क पर निर्धारित दर	500 वर्ग मीटर तक सड़क से 100 मीटर दूर निर्धारित दर	501 वर्ग मीटर से 1000 वर्ग मी. तक सड़क पर निर्धारित दर	501 वर्ग मीटर से 1000 वर्ग मीटर तक सड़क से 100 मी. दूर निर्धारित दर	निर्धारित दर	निर्धारित दर	निर्धारित दर	निर्धारित दर	निर्धारित दर
1	2	3	4	5	6	7	8	9	10	11	12	13
.	1. माजरा 2. ब्राह्मणवाला 3. धरतावाला 4. चिड़ोवाली	26	24	900	800	750	650	3000	65	50	35	35

1.	5. धर्मपुरडाण्डा 6. गोपीवाला 7. लोहारवाला 8. शाहपुर सन्तौर 9. राघढ़वाला 10. शाहनगर 11. चक शाहनगर 12. कण्डौली (केन्द्रीयदून) 13. कारगी ग्रान्ट 14. प्रेमपुर माफी 15. राजपुर माफी 16. अजबपुर कलां 16. अजबपुर खुर्द 18. केदारपुर 19. निरंजनपुर 20. कौलागढ़ मयचक भूड़ 21. चक डालनवाला 22. अधोईवाला 23. इन्दरपुर 24. धोरण खास 25. चीड़ोवाली 26. गढ़ी कैन्ट 27. थानी गांव 28. डुगाल गांव 29. वीरगिरवाली 30. कांवली शेष क्षेत्र (नगर निगम क्षेत्र की दरों के अन्तर्गत कांवली क्षेत्र में उल्लेखित कालोनी/मौहल्लों को छोड़कर) 31. प्रेम नगर

अर्द्धनगरीय क्षेत्र/विशिष्ट ग्राम

क्र. सं.	क्षेत्र/राजस्व ग्रामों के नाम	सिंचित दर प्रति हैक्टेयर (लाख रु. में)	असिंचित दर प्रति हैक्टेयर (लाख रु. में)	1000 वर्ग मीटर तक की भूमि के अन्तरण के लिए निर्धारित दरें (प्रति वर्ग मीटर)				व्यावसायिक दर प्रति वर्ग मीटर	दुकान/ रेस्टोरेंट/ कार्या. की प्रति वर्ग मीटर दर	अन्य व्यवसायिक प्रतिष्ठान की दर (प्रति वर्ग मीटर)		
										गोदाम	होटल/ गेस्ट हाउस/ बारातघर/ सिनेमा हॉल/ पैट्रोल पम्प	अस्पताल/ धर्मशाला/ मनोरंजन स्थल जो कि उ.प्र. वाणि. अधि. के अन्तर्गत है।
		निर्धारित दर	निर्धारित दर	500 वर्ग मीटर तक सड़क पर निर्धारित दर	500 वर्ग मीटर तक सड़क से 100 मीटर दूर निर्धारित दर	501 वर्ग मीटर से 1000 वर्ग मी. तक सड़क पर निर्धारित दर	501 वर्ग मीटर से 1000 वर्ग मीटर तक सड़क से 100 मी. दूर निर्धारित दर	निर्धारित दर	निर्धारित दर	निर्धारित दर	निर्धारित दर	निर्धारित दर
1	2	3	4	5	6	7	8	9	10	11	12	13
1.	1. भारूवाला ग्रान्ट 2. सेवलाखुर्द 3. मोहब्बेवाला 4. सौन्धोवाली 5. बन्जारे वाला 6. डांडा नूरीवाला 7. डांडा धोरण 8. डांडा लखौन्ड 9. ब्रह्मावाला 10. बाजावाला	25	21	850	750	700	650	2500	50	45	35	35

टिप्पणी–उक्त ग्रामों में जो भूमि सड़क से 350 मीटर के अंतर्गत स्थित होगी उस पर मुख्य सड़क की दरें प्रभावी होंगी। उक्त ग्रामों में जो भूमि नगर निगम क्षेत्रों के अन्तर्गत उल्लिखित मौहल्लों में स्थित होंगी उन पर नगर निगम के मौहल्लों की सुसंगत क्षेत्रान्तर्गत दरें लागू होंगी।

1	2	3	4	5	6	7	8	9	10	11	12	13
2.	1. जोहड़ी 2. सुन्दरवाला 3. आमवाला तरला 4. बद्रीपुर 5. नत्थनपुर 6. मोहकमपुर कलां 7. मोहकमपुर खुर्द 8. मालसी 9. सिनौला 10. दानियों का डाडां 11. आरकेडिया ग्रान्ट 12. गोरखपुर माफी 13. सेवला कलां 14. मोथरोंवाला 15. मेंहूवाला माफी 16. हरबंशवाला 17. नक्कावाला 18. हर्रावाला 19. कुवांवाला 20. माजरी माफी 21. मियांवाला 22. नवादा 23. बालावाला 24. लाडपुर 25. रायपुर 26. विजयपुर हाथीबड़कला 27. भारूवाला 28. तेलपुरा ग्रान्ट 29. चन्द्रबनी ग्रान्ट 30. कुठाल गांव 31. किरसाली 32. खेड़ा मानसिंह वाला 33. हरभजवाला 34. मिट्ठी भेडी 35. पित्थूवाला	25	21	850	750	700	650	2500	50	45	35	35

टिप्पणी—उपरोक्त सूची में वर्णित ग्रामों में मंसूरी रोड पर पड़ने वाली किसी भी क्षेत्रफल की भूमि के अन्तरण पर सड़क से 350 मीटर तक मुख्य सड़क की दरें प्रभावी होंगी।

ग्रामीण क्षेत्र

क्र. सं.	क्षेत्र/राजस्व ग्रामों के नाम	सिंचित दर प्रति हैक्टेयर (लाख रु. में)	असिंचित दर प्रति हैक्टेयर (लाख रु. में)	1000 वर्ग मीटर तक-की भूमि के अन्तरण के लिए निर्धारित दरें (प्रति वर्ग मीटर)				व्यावसायिक दर प्रति वर्ग मीटर	दुकान/ रेस्टोरेंट/ कार्या. की प्रति वर्ग मीटर दर	अन्य व्यवसायिक प्रतिष्ठान की दर (प्रति वर्ग मीटर)		
										गोदाम	होटल/ गेस्ट हाउस/ बारातघर/ सिनेमा हॉल/ पैट्रोल पम्प	अस्पताल/ धर्मशाला/ मनोरंजन स्थल जो कि उ.प्र. वाणि. अधि. के अन्तर्गत है।
		निर्धारित दर	निर्धारित दर	500 वर्ग मीटर तक सड़क पर निर्धारित दर	500 वर्ग मीटर तक सड़क से 100 मीटर दूर निर्धारित दर	501 वर्ग मीटर से 1000 वर्ग तक सड़क पर निर्धारित दर	501 वर्ग मीटर से 1000 वर्ग मीटर तक सड़क से 100 मी. दूर निर्धारित दर	निर्धारित दर	निर्धारित दर	निर्धारित दर	निर्धारित दर	निर्धारित दर
1	2	3	4	5	6	7	8	9	10	11	12	13
1.	1. हरिपुर 2. चकतुनवाला 3. नकरौन्दा 4. नथुवावाला	25	21	850	750	700	650	2000	50	35	25	25

टिप्पणी–उपरोक्त सभी ग्रामों में मुख्य हरिद्वार रोड पर हरिद्वार रोड के दोनों ओर 350 मीटर के अन्दर स्थित भूमि के अन्तरण पर हरिद्वार रोड की दर प्रभावी होंगी।

1	2	3	4	5	6	7	8	9	10	11	12	13
2.	1. ननूर खेड़ा 2. डांडा खुदानेवाला	25	21	850	750	700	650	2000	50	35	25	25
	3. अस्थल 4. आमवाला मझला 5. चांलग 6. कालागांव 7. गुजराडा नानसिंह 8. तरला नागल 9. आमवाला करनपुर											

टिप्पणी–उपरोक्त सभी ग्रामों में सहस्त्रधारा रोड पर सहस्त्रधारा रोड के दोनों ओर 350 मीटर के अन्दर स्थित भूमि के अन्तरण पर सहस्त्रधारा रोड की दरें प्रभावी होंगी।

1	2	3	4	5	6	7	8	9	10	11	12	13
3.	1. डोईवाला 2. डैशवाला 3. चान्दमारी	18	16	750	650	500	450	2000	50	35	25	25
	4. प्रेमनगर 5. कुड़कावाला 6. माधोवाला 7 लच्छीवाला 8. हंसूवाला 9. छदामीवाला 10. मारखम ग्रान्ट–I 11. मारखम ग्रान्ट– II 12. मिस्सरवाला कलां 13. मिस्सरवाला खुर्द 14. खैरी 15. खाता 16. धर्मूचक 17. तैलीवाला 18. बुल्लावाला 19. झबरावाला 20. दूधली 21. सिमलास ग्रान्ट 22. नागल ज्वालापुर 23. नागल बुलन्दावाला 24. फान्दूवाला 25. मौहम्मदपुर बड़कर्ली 26. किशनपुर 27. घिसरपड़ी											

टिप्पणी–उपरोक्त सभी ग्रामों में मुख्य हरिद्वार रोड पर हरिद्वार रोड के दोनों ओर 350 मीटर के अन्दर स्थित भूमि के अन्तरण पर हरिद्वार रोड की दर प्रभावी होगी।

डोईवाला टाउन एरिया क्षेत्र

क्र. सं.	क्षेत्र	सड़क से 50 मीटर तक आवासीय दर (प्रति वर्ग मीटर)	सड़क से 51 मीटर दूरी से 350 मीटर तक आवासीय दर (प्रति वर्ग मीटर)	व्यवसायिक दर (प्रति वर्ग मीटर)	रेस्टोरेंट/दुकान/कार्या. दर (प्रति वर्ग मीटर)	अन्य व्यवसायिक प्रतिष्ठान की दर (प्रति वर्ग मीटर) गोदाम	होटल/गेस्ट हाउस/बारातघर/ सिनेमाहाल/पैट्रोल पम्प	अस्पताल/धर्मशाला/ मनोरंजन स्थल/अन्य जो कि उ.प्र. वाणि. अधि. के अन्तर्गत है।
		निर्धारित दर	निर्धारित दर	निर्धारित दर	निर्धारित दर	निर्धारित दर	निर्धारित दर	निर्धारित दर
1	2	3	4	5	6	7	8	9
	1. देहरादून–डोईवाला–ऋषिकेश, शुगरमिल रोड तथा प्रेमनगर बाजार रोड पर	3000	2000	3000	65	50	64	64
	2. 350 मीटर के आगे	1800	1500	1800	50	40	35	35

देहरादून नगर निगम क्षेत्र की दरें

क्र. सं.	प्रमुख क्षेत्र	क्षेत्र	सड़क से 50 मीटर तक आवासीय दर (प्रति वर्ग मीटर)	सड़क से 50 मीटर आगे आवासीय दर (प्रति वर्ग मीटर)	व्यवसायिक दर (प्रति वर्ग मीटर)	रेस्टोरेंट/दुकान/कार्या. की दर (प्रति वर्ग मीटर)	अन्य व्यवसायिक प्रतिष्ठान की दर (प्रति वर्ग मीटर)		
							गोदाम	होटल/गेस्ट हाउस/बारात घर/ सिनेमाहाल/ पैट्रोल पम्प	अस्पताल/ धर्मशाला/ मनोरंजन स्थल जो कि उ.प्र. वाणि. अधि. के अन्तर्गत है।
			निर्धारित दर	निर्धारित दर	निर्धारित दर	निर्धारित दर	निर्धारित दर	निर्धारित दर	निर्धारित दर
1	2	3	4	5	6	7	8	9	10
1.	राजपुर क्षेत्र	1. राजपुर 2. ढाकपट्टी 3. किषनपुर 4. जाखन 5. प्रतीतपुर सन्तौर	2500	2200	3000	65	50	35	35
		1. वीरगिरवाली 2. घासमन्डी 3 बाडी गार्ड	2000	1800	3000	65	50	35	35
2.	आर्यनगर क्षेत्र	1. डी.एल. रोड 2. सेवक आश्रम रोड 3. विष्णु रोड 4. आर्यनगर 5. अधोईवाला 6. नदी रिस्पना रोड	2000	1800	3000	65	50	35	35
3.	मनसिंहवाला	1. ओल्ड सर्वे रोड 2. मानसिंहवाला 3. बंगाली लाइब्रेरी रोड 4. बंगाली मौहल्ला 5. डी.एल.वी. कालेज रोड 6. दिलाराम बाजार (राजपुर रोड से 350 मीटर छोड़कर)	2200	2000	3000	65	50	35	35
4.	ओल्ड डालनवाला	1. ओल्ड डालनवाला 2. करनपुर खास 3. गुरुद्वारा रोड 4. प्रिसिपिल रोड 5. सीमेन्ट रोड 6. नेगी रोड, 7. करनपुर मौहल्ला, 8. करनपुर बाजार	2500	2200	4000	85	60	35	35
5.	न्यू डालनवाला	1. म्यूनिसिपल रोड 2. इन्दर रोड 3. प्रीतम रोड 4. मोहिनी रोड 5. बलबीर रोड 6. तेग	3200	3000	4000	85	60	35	35

1	2	3	4	5	6	7	8	9	10
		बहादुर रोड 7. सरकुलर रोड 8. नेहरू रोड 9. कर्जन रोड 10. लक्ष्मी रोड 11. चन्दर रोड 12. नेमी रोड (इसके अतिरिक्त डालन वाला क्षेत्र में पड़ने वाला समस्त सड़कें तथा मौजा धर्मपुर का नहर/नालें से उत्तर पूर्व का डालन वाला में पड़ने वाला समस्त क्षेत्र), 13. उग्र रोड, 14. डिक रोड							
		डालनवाला नदी क्षेत्र में नदी के किनारें से 150 मीटर तक तथा एम.डी.डी.ए. कालोनी	2400	2200	4000	85	60	35	35
6.	फालतू लाईन	1. पटेल रोड 2. न्यू रोड 3. कान्वैन्ट रोड 4. पन्त रोड 5. नरदेव शास्त्री मार्ग, 6. इन्दिरा मार्केट	4000	3000	4000	85	60	35	35
		1. कलालों वाली गलीं	3000	2800	4000	85	60	35	35
7.	धर्मपुर क्षेत्र	1. धर्मपुर 2. नेहरू कालोनी 3. आराघर (उक्त तीनों क्षेत्रों के हरिद्वार रोड के दोनों तरफ 350 मीटर छोड़कर तथा डालनवाला क्षेत्र में आच्छादित क्षेत्र को छोड़कर)	2600	2400	4000	85	60	35	35
		शास्त्रीनगर (समस्त लेन)	2000	1800	3000	65	50	35	35
8.	रेसकोर्स	रेसकार्स	3200	3000	4000	85	60	35	35
9.	देहराखास	1. रैस्टकैम्प 2. चन्दर नगर 3. त्यागी रोड 4. सन्त विनोबा भावे मार्ग (नाले से पूर्व)	2200	2000	3500	85	60	35	35
		1. सन्त विनोबा भावे मार्ग नाले के बाद 2. देहराखास अन्य शेष क्षेत्र	1800	1600	3000	65	50	35	35
		1. रीठामन्डी 2. सिंगलमन्डी 3. मुस्लिम कालोनी 4. पथरी बाग 5. भन्डारी बाग समस्त ब्लाक 6. रामनगर 7. लक्खीबाग (क्रमांक 6 एवं 7 के वह क्षेत्र जो कि	1800	1600	3000	85	60	35	35

1	2	3	4	5	6	7	8	9	10
		सहारनपुर रोड के दोनों तरफ 350 मीटर छोड़कर स्थित है)							
10.	लक्ष्मण चौक	1. महन्त रोड 2. मालवीय रोड 3. नेशनल रोड 4. पी.डी. टन्डन रोड 5. विचारानन्द मार्ग 6. पार्क रोड 7. केशव रोड 8. सरस्वती सोनी मार्ग 9. गुरू रोड	2400	2200	4000	85	60	35	35
	पटेल नगर	1. पार्क रोड एवं गुरु रोड तथा सरस्वती सोनी नार्ग (तीनों सड़कों) का नदी क्षेत्र 2. ईस्ट पटेल नगर 3. वैस्ट पटेल नगर 4. शिवलोक कालोनी, 5. गांधी ग्राम 6. नेहरू नगर (उक्त क्षेत्रों के समस्त क्षेत्र जो कि कांवली रोड व सहारनपुर रोड से 350 मीटर छोड़कर स्थित है)	2200	2000	4000	85	60	35	35
12.	इन्द्रेश नगर	1. इन्द्रेश नगर 2. जटिया मौहल्ला	1500	1300	3000	65	50	35	35
13.	शिवाजी क्षेत्र	1. छबील बाग 2. खदरी मौहल्ला 3. बिन्दाल रोड	2000	1800	3000	65	50	35	35
		1. बिन्दाल रोड का निचला भाग नदी क्षेत्र	1600	1400	3000	65	50	35	35
14.	खुडबुडा क्षेत्र	1. खुड़बुड़ा समस्त ब्लाक 2. अजीत प्रसाद मार्ग 3. अखाडा मौहल्ला 4. रामेश्वर मौहल्ला 5. आनन्द चौक 6. मालियान मौहल्ला 7. नेताजी मौहल्ला	2300	2100	3500	65	50	35	35
15.	हकीकत राय नगर	1. हकीकत राय नगर 2. मन्नू गंज 3. डाडींपुर मौहल्ला 4. गुजराती मौहल्ला	2000	1800	4000	65	50	35	35
16.	लूनिया मौहल्ला	1. लूनिया मौहल्ला	2400	2200	4000	65	50	35	35
17.	नारायण दत्त डंगवाल मार्ग	1. डंगवाल मार्ग 2. बकरालवाला 3. डोमालवाला 4. परसोलीवाला 5. अहीरमन्डी 6. पत्थरियापीर	1800	1600	3500	65	50	35	35
			1800	1600	3000	65	50	35	35

1	2	3	4	5	6	7	8	9	10
18.	चुक्खुवाला क्षेत्र	1. इन्दिरा कालोनी 2. चुक्खुवाला नई बस्ती 3. चुक्खूवाला 4. मित्रलोक कालोनी 5. ओंकार रोड	1800	1600	3000	65	50	35	35
19.	हाथीबडकला क्षेत्र	1. सालावाला 2. विजय कालोनी 3. हाथीबड़कला	1000	–	2000	65	50	35	35
		1. हाथी बड़कला नदी क्षेत्र	2000	1800	4000	85	60	35	35
20.	राजेन्द्र नगर	1. राजेन्द्र नगर 2. बल्लूपुर 3. कालागढ़ 4. गढ़ी उददीवाला 5. दीपलोक 6. नरेन्द्र विहार	2000	1800	4000	85	60	35	35
		1. सैययद मौहल्ला 2. किशन नगर 3. पंडितवाड़ी	2000	1800	4000	85	60	35	35
		1. कौलागढ़ मयचक	2000	1800	4000	85	60	35	35
21.	यमुना कालोनी	1. यमुना कालोनी 2. गोविन्दगढ़ ईदगाह 4. कुम्हार मन्डी, 5. टीचर्स कालोनी	2000	1800	4000	85	60	35	35
22.	कांवली क्षेत्र	1. इन्द्रानगर कालोनी 2. बसंत विहार	2300	2100	4000	85	60	35	35
		1. मोहितनगर 2. आशीर्वाद एन्कलेव 3. वन विहार 4. विजय पार्क 5. विजय पार्क एक्सटेंशन, 6. दत्ता एनक्लेव 7. इंजीनियर्स एनक्लेव समस्त फेज 8. हिल व्यू कालोनी 9. उत्सव विहार 10. कालिन्दी एनक्लेव 11. दुर्गा विहार 12. विवेक विहार 13. नेहरूपुरम् एनक्लेव 14. संगम विहार 15. अंकित पुरम एनक्लेव 16. नर्मदा एनक्लेव 17. मेघा एनक्लेव 18. पुष्पांजलि एनक्लेव 19. शाम्भवी लोक 20. काली मन्दिर एनक्लेव 21. द्रोणपुरी एनक्लेव 22. अलकनन्दा एनक्लेव 23. मिलन विहार	2000	1800	4000	85	60	35	35

1	2	3	4	5	6	7	8	9	10
		एनक्लेव 24. शक्ति एनक्लेव 25. द्वारिकापुरी 26. व्योमप्रस्थ एनक्लेव 27. प्रियदर्शनी एनक्लेव 28. शिवालिकपुरम 29. गढ़वाला कालोनी 30. जनकपुरी एनक्लेव 31. अशोक विहार							

देहरादून नगर के प्रमुख मार्गों पर स्थित क्षेत्रों की दरें

क्र. सं.	मोटर मार्ग	क्षेत्र	सड़क से 50 मीटर तक आवासीय दर (प्रति वर्ग मीटर)	सड़क से 51 मीटर से 350 मीटर तक आवासीय दर (प्रति वर्ग मीटर)	व्यवसायिक दर (प्रति वर्ग मीटर)	दुकान/रेस्टोरेंट कार्या. की दर (प्रति वर्ग मीटर)	अन्य व्यवसायिक प्रतिष्ठान की दर (प्रति वर्ग मीटर)		
							गोदाम	होटल/गेस्ट हाउस/ बारातघर/ सिनेमाहाल/ पैट्रोल पम्प	अस्पताल/ धर्मशाला/मनोरंजन स्थल/अन्य जो कि उ.प्र. वाणि. अधि. के अन्तर्गत है।
			निर्धारित दर	निर्धारित दर	निर्धारित दर	निर्धारित दर	निर्धारित दर	निर्धारित दर	निर्धारित दर
1	2	3	4	5	6	7	8	9	10
1.	राजपुर रोड	घंटाघर से मसूरी बाईपास तक	6500	4500	6500	120	100	65	64
		एस्से हॉल, प्रिंसिपल लक्ष्मण प्रसाद रोड	6500	4500	6500	120	100	64	64
		मसूरी बाईपास से राजपुर तक	5000	3500	5000	86	60	64	64
		मसूरी बाईपास से मालसी डीयर पार्क तक	4000	2700	4000	65	50	64	64
		मालसी डीयर पार्क से कुठालगेट तक	3500	2500	3500	44	35	64	64
		कुठाल गेट से राजपुर तक	3500	2500	3500	44	35	64	64
2.	ईस्ट कैनाल रोड	ई.सी. रोड, सुभाष रोड, क्रास रोड, न्यू सर्वे रोड	6000	4500	6000	108	35	64	64
3.	रायपुर रोड	सर्वे चौक से रिस्पना पुल तक	4000	3000	4000	86	60	64	64
		रिस्पना पुल से रायपुर तक	3000	2500	3000	85	65	64	64

1	2	3	4	5	6	7	8	9	10
		सहस्त्रधारा चौक से सहस्त्रधारा रोड पर छतरी तक	2500	2000	2000	60	60	60	60
4.	न्यू कैन्ट रोड	दिलाराम बाजार से लोहे के पुल तक	5000	2100	5000	85	65	64	64
		लोहे के पुल से गढ़ी डाकरा बाजार तक	3000	2500	3000	65	50	64	64
5.	चकराता रोड	घंटाघर से बिन्दाल पुल तक	6000	4500	6000	108	90	64	64
		बिदाल पुल से किशन नगर चौक तक	5000	3100	5000	86	60	64	64
		किशन नगर चौक से बल्लूपुर चौराहे तक	5000	3100	5000	86	60	64	64
		बल्लूपुर चौक से प्रेमनगर बस स्टैन्ड तक	4000	3000	4000	65	50	64	64
		प्रेमनगर बस स्टैन्ड से नदी तक	2000	1300	3000	65	50	64	64
		चुक्खुवाला मार्ग, नारी शिल्प मन्दिर मार्ग, रामपुर मण्डी रोड (चकराता रोड छोड़कर)	5000	3100	5000	86	60	64	64
6.	कौलागढ़ रोड	किशन नगर चौक से सैन्ट्रल स्कूल तक	4000	2900	4000	86	60	64	64
		सैन्ट्रल स्कूल से आगे कौलागढ़ की सीमा तक	3500	2500	3500	65	50	64	64
7.	सहारनपुर रोड	सहारनपुर चौक से बिन्दाल पुल तक	6000	4400	6000	108	90	64	64
		बिन्दाल पुल से टर्नर रोड तक	5000	3100	5000	86	60	64	64
		टर्नर रोड से सुभाष नगर चौक तक	4000	3000	4000	65	50	64	64
		टर्नर रोड, पोस्ट ऑफिस रोड	2000	1800	2000	65	50	64	64
		सुभाष नगर चौक से आगे वन क्षेत्र तक	2000	1800	2000	65	50	64	54
8.	कांवली रोड	सहारनपुर चौक से नदी तक	5000	3500	5000	86	60	64	64
		नदी से आगे बल्लीवाला चौक तक	4000	3000	4000	65	50	64	64
9.	हरिद्वार रोड	प्रिंस चौक से रिस्पना पुल तक	6000	4500	6000	108	90	64	64
		रिस्पना पुल से हर्रावाला तक	4000	3500	4000	86	65	64	64
		हर्रावाला से कुआंवाला तक	2400	2200	2400	65	50	64	64
		कुआंवाला से डोईवाला तक	2400	2200	2400	65	50	64	64
		बाईपास सहारनपुर रोड से हरिद्वार रोड तक	2200	2000	2200	86	65	64	64
		मोथरो वाला रोड	2400	2200	2400	65	50	64	64
10.	गांधी रोड	घंटाघर से रेलवे स्टेशन तक	6500	–	6500	110	90	64	64

1	2	3	4	5	6	7	8	9	10
		रेलवे स्टेशन से सहारनपुर रोड पर भण्डारी बाग जाने वाली सड़क तक	6500	–	6500	110	90	64	64
		आढ़त बाजार	10000	–	10000	120	100	64	64
11.	तिलकरोड	तिलक रोड	5000	–	5000	86	65	64	64
12.	अंसारीरोड	अंसारी रोड	5000	–	5000	86	65	64	64
13.	पल्टन बाजार	घंटाघर से कोतवाली तक	11000	–	11000	129	100	64	64
14.	धामावाला/पीपलमन्डी	कोतवाली से लक्खी बाग चौकी तक	10000	–	10000	120	100	64	64
		दर्शनी गेट/आढ़त बाजार	10000	–	10000	120	100	64	64
		1. धामावाला 2. राजा रोड 3. अजमल खान रोड 4. डिस्पैन्सरी रोड 5. सरनीमल बाजार 6. मोती बाजार 7. पीपल मन्डी 8. हनुमान चौक 9. दर्शनी गेट 10. झन्डा मौहल्ला/बाजार 11. बाबू गंज 12. घोसी गली	6500	6000	6500	110	90	64	64
		1. जैन कालोनी	6000	4500	6000	65	50	35	35
15.	कालीदास रोड	काली दास रोड	3000	2800	3000	65	50	64	64
16.	नैशिवला रोड	राजुपर रोड से काम्बोज स्वीट शाप तक	4000	3000	4000	65	50	64	64
		काम्बोज स्वीट शाप से छोटी बिन्दाल नदी तक	3500	2800	3500	65	50	64	64
17.	जनरल महादेव सिंह रोड	चकराता रोड क्रासिंग से बल्लीवाला चौक कांवली रोड क्रासिंग तक	5000	4000	5000	85	65	64	64
		बल्लीवाला चौक कांवली रोड क्रासिंग से सहारनपुर रोड तक	5000	4000	5000	85	65	64	64
		इन्द्रापुरम्	4000	3000	4000	65	50	64	64

B. Schedules/Questionnaire

1. Questionnaire for Tehsil Office
2. Name of the Tehsil.......................................
3. Total number of Villages...................................
4. Total Area...
5. Number of Development Blocks............................
6. Total number of Land Acquisitions in Last 10 Years:

Sl. No.	*Name of Block*	*Total No. of Cases*	*Purpose*	*Achieved or Not*

7. Total Number of Mutations:..........................

Name of the Block	*No. of Mutations*	*Year*	*Method of Transfer*

8. Total Number of Khata Dharak/Land Holders:

Sl. No.	*Name of Block*	*Area*	*Social Categories*			
			SC	*ST*	*OBC*	*Others*

9. Cases of Change in Topography:

Name of Block	*Change in Topography*	*Area*	*Original Topography*	*Remarks*

10. Change in Nature of Land:

Name of Block	*Change in Nature of Land*	*Area*	*Original Function of Land*	*Remarks*

11. Reasons for Demand of Land in the Land Market in the Tehsil:

Category	*Reasons*							
	Roads & Railways	*Government Building*	*Small Industries*	*Big Industries*	*SEZ*	*Residence Purpose*	*Real Estate*	*Others (Please specify)*
Agricultural Land								
Village Community/ Civil Land								
Waste Land								
Civil Forest Land								
Reserve Forest Land								

12. Land sale or purchase by the non-resident of the village is helpful for village economy?

 1. Yes......... 2. No...........

 If yes, then in what way it is beneficial?

13. After creation of Uttarakhand State and Dehradun made as its capital, do you feel that rapid urbanization and industrialization has been increased in this area? Your comments?

    ....
    ....
    ....
    ....

14. Do you think that the pressure on land has been increased due to the urbanization and industrialization? Your opinion please?

    ....
    ....
    ....
    ....

15. Do you feel that land sale for other than agriculture purpose has been increased?
16. Do you observe that pressure on fertile agriculture land is more in comparison with waste land or barren land?
17. Forest land is increased or decreased in your area after formation of Uttarakhand State?

    ....
    ....
    ....
    ....

18. Is land available easily for land market? How State government is facilitating in favour of industrialization?

    ....
    ....
    ....
    ....

19. Is change in land-use through land markets for non-agricultural purpose is beneficial for the economy of the common people?

    ....
    ....
    ....
    ....

20. Is it purely serving industrialists and other market forces and enhancing government revenue?
    ....
    ....
    ....
    ....

    Comments of Official:
    (i) ....
    (ii) ....
    (iii) ....

Signature of the Official..............................

Name of the Official..................................

2. Questionnaire for Patwari

1. Name of the Patwari...
2. No. of Villages under charge................................
3. Total Area...
4. Development Block ....
5. District ....
6. State ....
7. Educational Qualification......................................
8. Total Number of Plots Village-wise:

Sl. No.	*Name of Village*	*Total No. of Plots*	*Remarks*

9. Total Number of Khatadharak/Land Holders:

Sl. No.	*Name of Village*	*Name of biggest Khatadharak/Land Holder*	*Area of Plot*	*Name of Smallest Khatadharak/Land Holder*	*Area of Plot*	*Remarks*

10. Occupations in your area:

Mode of Occupation	No. of Persons Involved	Area engaged	Income
Agriculture			
Dairy			
Fishery			
Poultry			
Real Estate			
Factory			
Other Business			
Service			

11. Total Agricultural Land:
    Irrigated:..........................
    Unirrigated.......................

Name of Village	Agricultural Land (Area)	Irrigated Land (Area)	Unirrigated Land (Area)

12. Cases of Sale and Purchase of any land in last 5 years:

Name of Village	Cases of Land Sale	Area	Cases of Land Purchase	Area

13. Total Number of Mutations:..........................

Name of the Village	*No. of Mutations*	*Year*	*Method of Transfer*

14. Category of Law:

Land Transactions	*Category of Law*	*Rules needs modification (Your opinion)*
Mutations		
Land Sale		
Land Purchase		
Land Acquired		
Land Lease/Rent		
Change in Land Use		

15. After creation of Uttarakhand State and Dehradun made as its capital, do you feel that rapid urbanization and industrialization has been increased in this area? Your comments?

    ....

    ....

    ....

16. Do you think that the pressure on land has been increased due to the urbanization and industrialization? Your opinion please?

    ....

    ....

    ....

    ....

17. Do you feel that land sale for other than agriculture purpose has been increased?

    ....

    ....

    ....

    ....

18. Do you observe that pressure on fertile agriculture land is more in comparison with waste land or barren land?

 ....

 ....

 ....

 ....

19. Forest land is increased or decreased in your area after formation of Uttarakhand State?

 ....

 ....

 ....

 ....

20. Is land available easily for land market? How State government is facilitating in favour of industrialization?
21. Is change in land-use through land markets for non-agricultural purpose is beneficial for the economy of the common people?

 ....

 ....

 ....

 ....

22. Is it purely serving industrialists and other market forces and enhancing government revenue?

 ....

 ....

 ....

 ....

 Comments of Official:

 (i) ....

 (ii) ....

 (iii) ....

Signature of the Official..

Name of the Official..

3. Schedule for Villager

1. Name of the farmer............................
2. Village..
3. Development Block ....
4. District ....
5. State..
6. Caste..
7. Education..
8. Size of family: Adult male... Adult female children.........
9. No. of persons engaged in agriculture and allied (poultry/ dairy/fishery, etc.) activities ....
10. No. of persons engaged in business ....
11. No. of persons engaged in service ....
12. Annual income from agriculture
13. (i) Other allied (poultry/dairy/fishery, etc.) activities if any ....
    Annual Income from that activities (Rs).................
    (ii) Other trade and business of any family member
    Annual income from ....(Rs)
    (iii) Annual income from Service (of any family member)..........................(Rs)
14. Size of landholding: Total .... (acres)
    Distance from main road.......................
    (i) Irrigated .... (acres)
    (ii) Unirrigated ....(acres)
    Category of Land

Category of Land	*Area (acres)*	*Land engaged in*

A. Have you sold or purchased any land in last few years?
   1. Yes......   2. No........

B. What was the size of land?
   (a) Below 1 Acre (b) 1 to 5 Acres
   (c) Above 5 Acres

C. Please tell how it is beneficial to you?: (Tick mark ✓)
   1. Social status is improved.
   2. Family income is increased by doing farming on it.
   3. Land is for business purpose (income gain either from sale or purchase).
   4. Any other reason.

15. Any land in your name is acquired by the government? If yes, then kindly tell:
    (a) What was the size of land:............
    (b) What was the compensation received...............
    (c) When it was acquired.................................
    (d) For what purpose....................................
16. Family Structure: (Tick mark ✓)
    A. Nuclear Family
    B. Combined Family

    How many members of your family have gone outside in search of livelihood...........

    Reasons for that: (Tick mark ✓)
    1. Agriculture is not sufficient
    2. Land is not sufficient
    3. Any other reason
17. Have you sold any land in last ten years? (Tick mark ✓)
    A. Yes B. No

    If yes, please tell for what purpose: (Tick mark ✓)
    1. House construction
    2. Marriage
    3. To start business
    4. Cottage industry/Poultry
    5. Any other reason......................
18. Have you purchased any land in last ten years? (Tick mark ✓)
    A. Yes B. No

If yes, please tell for what purpose: (Tick mark ✓)

1. House construction
2. Marriage
3. To start business
4. Agriculture
5. Cottage industry/Poultry
6. Any other reason........................

19. Role of Gram Pradhan/Sarpanch in the land sale or purchase ? (Tick mark ✓)

    A. Yes B. No

    What type of Role? ....

    ....

20. Role of Bank in the land sale or purchase:(Tick mark ✓)

    A. Yes B. No

    What type of Role? ....

21. What type of documents you need for the land sale or purchase:

    (i) ....

    (ii) ....

    (iii) ....

22. What specific problem/s did you face during initiating the land sale or purchase:

    (i) ....

    (ii) ....

23. Has somebody mediated it?

    1. Yes....... 2. No......... (Tick mark ✓)

24. Have you paid to the mediator? If yes, how much.................................

25. What is the selling (market) price/cost of the land (per acre)?

26. Rental value of the agricultural land: .... (Rs./acre)

27. Whether your family income has increased by giving your land on rent?

    1. Yes....... 2. No.........(Tick mark ✓)

28. Did you take loan from anyone on land ?

1. Yes....... 2. No.........(Tick mark ✓)
   If yes,
   (a) From any mahajan/landlord of the village/company/ bank/or government institution
   (b) If from bank/government institution, name of the bank/ government institution............................ ......
   (c) the amount of loan (Rs)
   (d) Purpose of the loan ....
   (e) Rate of interest .... (%)
   (f) Whether repaid on time? Yes / No

29. Have ever been your land is mortgaged by the Bank?
    1. Yes....... 2. No...... (Tick mark ✓)

30. Do you think that taking loan on land has increased land sale or purchase?
    1. Yes....... 2. No......... (Tick mark ✓)
    How?
    (a) ....
    (b) ....
    (c) ....

31. Production on Agricultural Land:

Land Quantity (In Acres)	*Production under Ravi season (In quintals)*			*Production under Kharif season (In quintals)*			*Production under Jayad season (In quintals)*		
	Crop	*Crop*	*Crop*	*Crop*	*Crop*	*Crop*	*Crop*	*Crop*	*Crop*

Comments of Research Investigator:
(i) ....
(ii) ....
(iii) ....

Signature of the Research Investigator..............................

Name of the Research Investigator.................................

4. Questionnaire for District Revenue Office

1. Name of the District..................................
2. Total number of Villages..................................
3. Total Area..
4. Number of Development Blocks........................
5. Number of Tehsils....................................
6. Total number of Land Acquisitions in Last 10 Years:

Sl. No.	Name of Tehsil	Total No. of Cases	Purpose	Achieved or Not

7. Total number of Khatadharak/Land Holders:

Sl. No.	Name of Tehsil	Area	Area	Name of Village	Area	Remarks

8. Total Agricultural Land:
   Irrigated:............................
   Unirrigated........................

Name of Tehsil	Agricultural Land (Area)	Irrigated Land (Area)	Unirrigated Land (Area)

9. Cases of Change in Topography:

Name of Tehsil/ Village	Change in Topography	Area	Original Topography	Remarks

10. After creation of Uttarakhand State and Dehradun made as its capital, do you feel that rapid urbanization and industrialization has been increased in this area? Your comments?

    ....

    ....

    ....

    ....

11. Do you think that the pressure on land has been increased due to the urbanization and industrialization? Your opinion please?

    ....

    ....

    ....

    ....

12. Do you feel that land sale for other than agriculture purpose has been increased?

    ....

    ....

    ....

    ....

13. Do you observe that pressure on fertile agriculture land is more in comparison with waste land or barren land?

    ....

    ....

....

....

....

14. Forest land is increased or decreased in your area after formation of Uttarakhand State?

....

....

....

....

15. Is land available easily for land market? How State government is facilitating in favour of industrialization?

....

....

....

....

16. Is change in land-use through land markets for non-agricultural purpose is beneficial for the economy of the common people?

....

....

....

....

17. Is it purely serving industrialists and other market forces and enhancing government revenue?

....

....

....

....

Comments of Official

(i) ....

(ii) ....

(iii) ....

Signature of the Official..

Name of the Official..

Bibliography

Acheson, J.M., (1989): Economic Anthropology and the Management of Common Property Resources. *Economic Anthropology*, Stanford University Press, Stanford.

Anderson, E.N., (1987): *A Malaysian Tragedy of the Commons, The Questions of the Common*, University of Arizona Press, Tucson.

Arnold, J.E.M. and W.C. Stewart. (1991): Common Property Resource Management in India, *Tropical Forestry Papers*, No. 24, Oxford Forestry Institute, University of Oxford.

Atkinson, E.T., (1882): *The Himalaya Gazetteers*, Vol. II, Part I, Cosmo Publications, Delhi.

Babette, W. (2008). The Dynamics of Peri-Urban Land Markets in Sub-Saharan Africa: Adherence to the Virtue of Common Property *vs*. Quest for Individual Gain. Erdkunde.

Berkes, F., (1987): Common Property Resource Management and Cree Indian Fisheries in Sub-arctic Canada, *The Question of the Commons*, University of Arizona Press, Tucson.

——, Feeny, D. and *et al*., (1989): The Benefits of the Commons, *Nature*, 340: 91-93.

——, (1977): Fishing resource use in a Sub-arctic Indian community, *Human Ecology*, 5:289-307.

——, (1985): Fishermen and the 'Tragedy of the commons'. *Environmental Conservation*, Belhava, London.

——, (1989): *Common Property Resource: Ecology and Community based Sustainable Development*, Belhaven, London.

Berreman, G.D., (1993): *Hindus of the Himalayas: Ethnography and Change*, Oxford University Press, Bombay.

Birner, R. (1999). *The Role of Livestock in Agriculture Development*. Theoretical Approaches and their Application in the Case of Sri Lanka.

Biswas, S.K., (1996): *Central Himalayan Panorama*, D.K. Publishers Distributors (P) Ltd., New Delhi: 90.

Bose, S.C., (1972): *Geography of the Himalaya*, National Book Trust, New Delhi.

——, (1976): *Geography of the Himalayas*, National Book Trust, New Delhi.

Dale, Peter, R.M. (2013). *Land Markets and the Modern Economy*. Retrieved from www.rics.org.

FAO. (2002). *Land Tenure and Rural Developmant*. In FAO, FAO Land Tenure Studies 3. Rome.

Foncier, M. (1999). Tax and economic terms in land. (p.114) Retrieved from http://www.fao.org/docrep/005/X2038E/x2038e09.htm.

Gupta, Kshama (2010): Unprecedented growth of Dehradun Urban Area: A Spatio-Temporal Analysis, Indian Institute of Remote Sensing, Dehradun, India.

Janssen, Larry, Jack B. Davis and Sarah Adams Inkoom. (2015). South Dakota Agricultural Land Market Trends, 1991-2015: Results from the 2015 SDSU South Dakota Farm Real Estate Survey. Retrieved from https://books.google.co.in/books

Kihato, C.W., and Momoniat, L.G. (2010). *Urban Land Markets: Economic Concepts and Tools for engaging in Africa*. (p. 14) Retrieved from www.unhabitat.org/pmss/getElectronicVersion.

Kivell, P.T. (1987). *Public Ownership of Urban Land*. Keele, Staffordshire: Department of Geography, University of Keele, Keele, Staffordshire.

Kuldeep, Tiwari and Khanduri, Kamlesh, (2010): *Land Use/Land cover change detection in Dun valley (Dehradun Tehsil)*, Uttarakhand: Using GIS & Remote Sensing Technique.

Mahoney, Robert, P.D. (2007, July). *Land Markets — Why are They Required and How Will They Develop*. (p.1) retrieved from http://www.fig.net/pub/monthly_articles/july_2007/july_2007_mahoney_dale_mclaren.pdf.

Oyebanji, O-Oyeyinka. (2013). *State of Urban Youth report 2012-2013*. Retrieved from www.unhabitat.org/pmss/getElectronicVersion.

Sexton, Robert L. . (2015). *Exploring Economics*. Retrieved from https://books.google.co.in.

Simpson, S.R.. (1976). Land Law and Registration. In S.R. Simpson, *Land Law and Registration* (p. 3). Cambridge: Cambridge University Press.

Singh, V.V. (2004) "The Jaunsari: A study in Nature-Man-Spirit complex of a Himalayan Tribe," *Tribe*, Udaipur, Vol. 34 No. 1-4.

—— (2002) "Usufruct in Uttaranchal: Some findings", *Yojana*, Planning Commission, New Delhi, Vol. 46.

—— (2007) "Common Property Resources Utilization and its Effect on the Lifestyle of the Jaunsari Tribals in the Himalayan Geo-Physical Setting. Un-Published Ph.D. Thesis. Department of Sociology, University of Lucknow, Lucknow.

—— (2010) "Socio-economic State Paper of Orrisa," *Socio-Economic Profile of Rural India*, Vol. IV. Series II, Eastern India, Concept Publishing Company (P) Ltd., New Delhi.

Vermeulen, P.C. (2009). *Land Markets and their Regulation: The Welfare Economics of Planning*. (pp. 121) Retrieved from http://eprints.lse.ac.uk/30787/1/Land_markets_and_their_regulation_(LSERO_version).pdf.

Wehrmann, B. (2005 & 2006). *Urban and Peri-Urban Land Conflicts in Developing Countries*. Research Reports on Urban and Regional Geography; Cadastre in Itself won't Solve the Problem: The Role of Institutional Change and Psychological Motivation in Land Conflicts. Berlin.

World Bank (2016). *Doing Business 2016: Measuring Regulatory Quality and Efficiency*.

World Bank (2016). *Global Monitoring Report* 2015/2016.

Index

Agriculture/Agricultural
and allied sector economy, 16
diversification, 57
employment, 33
labour, 56
occupations, 56
operations, 18
production systems, 21
purposes, 35, 43
Agro-climatic zones, 6
Allied Sector Economy, 16
Allied sectors, 5, 16
Alluvial soils, 12
Alpine zone, 12
Altitudinal range, 12
Animal husbandry, 20, 57
Animal rearing, 16
Arable land, 16–17, 23–24
Artisans groups, 17

Balance of power, 3
Bania middleman, 21
Barren land, 73, 77, 83
Bio- fertilizer application, 22
Brahmin and Rajput caste, 22
Building Operations Act, 30

Capital investment, 56
Cascading effect, 59
Caste differentiation, 21
Caste sub-groups, 17
Civilized hills, 24
Climatic regions, 8–9
Climatic seasons, 10
Commercial prices, 50
Common land tenure system, 22
Community
control, 25
earnings, 51
level society, 54
Complementary institutions, 3
Conflict resolution, 26
Construction sector, 31
Consumer rights organizations, 3
Contract agreement, 43
Copy documents, 6
Credit mobilization, 58
Cultural and Social recognition, 52
Curtail amassment, 4
Customary laws, 54
traditional, 53

Development
planners, 28–29
socio-economic, 59
sustainable, 53
Disadvantaged groups, inclusion of, 53
Domestic animals productivity, 22
Domestication, 24
Dynamics of Land Markets, 28

Earning occupations, 56

Ecological Sustainability, 27
Economic
    groups, 56
    reform, 57
    Sectors, 59
Employment opportunities, 25
Environmental sustainability, 59
Equal rights, 53

Factors
    critical, 2
    demotivating, 45
    economic, 30, 51
Fair trade, 21
Fauna, 13, 39
Fertile agriculture land, 73, 77, 83
Field Survey, 15, 17–19, 35–36, 38, 41, 44–47
Financial mechanisms, 3
Flash floods, 10
Food procurement system, 23
Food production system, 23, 56
Forest
    boundaries, 24
    contribution of resources, 5, 25
    department, 30
    land, 23, 52, 73, 77, 84
    management, 26
    products, 23
    region, 12
    resources, 23–27
    rich, 12
Forest Land Resource Conservation, 27
Forestry, 16
Formal systems, 33
Free fodder, 24
Fruit gardens, 21

Ganges, 7
Geomorphology, 8
Government revenue, 48
Gram Panchayat, 47
Gram Pradhan, 47
Gram Sabha, 47
Grazing space, 25
Greater Himalaya, 8
Group survival, 26

Hill regions, 53
Hilly districts, 16
Hilly pastures, 24
Hilly tracts, 32
Honey bees, 21
Hydrological resources, 8

Illegal colonization, 52
Illegal transactions, 52
Immovable land property, 48
Indian Military Academy, 29
Indian Stamp Act, 48
Industries, tea, 30
Informal agreement, 33
Informal land markets, 52
Information and Communication Technology (ICT), 39
Institutional management, 6
Insurance facilities, 57
International Development, 52
International trade organizations, 3
Investment power, 33
Irrigated
    facilities, 18
    terraces, 18

Kharif pulses, 19

Labour market opportunities, 58
Land administration system, 2, 45
Land administration, 45, 59
Land ceiling provision, 4
Land clash, 57
Land contracts, 56
Land developmental activities, 30
Land holders, 22
Land markets, 2, 28–29, 31–39, 41–48, 51–59, 72–73, 77, 84
    activities, 30, 34, 41, 48, 53–54, 58

efficiency, 43
systems, 28, 56
transaction, 38, 45, 48
Land masses, 4
Land Mutations, 45–46
Land owner communities, 52, 56–57
Land reforms, 4
Land registrar, 45
Land registration system, 2
Land registrations, 45, 49, 58
Land rental, 4, 34, 54, 56
Land rights, 2, 42, 52, 58
Land tenures, 21, 33, 52
and rural development, 3
transactions, 43
Land titling, 33, 45, 48, 58
Land transactions, 1–2, 5, 34, 42–44, 47–48, 58
Land-use
changes, 31, 47, 52
dynamics of, 5, 30
Legitimate landowner, 45
Lesser Himalaya, 8
Literacy rate, 13
Localized communities, 3

Mahajan/landlord, 81
Male-Female ratio, 15
Mammals, 13
Management of land, 52
Marginalized social groups, 59
Market players, 2, 46
Material cultures, 1
Mean Sea Level (MSL), 7–9, 11–12
Medicinal plants, 22, 24
Medicines, 25
Moderate fertility, 11
Monitoring mechanism, 5
Monsoon zone, 18
Mortgage agreement, 34
Mountain slopes, 16
Mountainous region of district, 18, 20–21
Multiple revenue village, 47
Mussoorie hills, 20
Mussoorie-Dehradun Development Authority (MDDA), 46

National land policy, 2
Non-agricultural purposes, 5, 46, 73, 77, 84
Non-agriculture sectors, 4
Non-judicial Stamp Duties, 48

Palearctic region, 13
Pastoral economy, 24
Pasture land, 7, 23
Patanjali Products, 22
Physiography, 7
Political economy, 3
Premium prices, 21
Pre-monsoon showers, 10
Price mechanism, 2
Primary sector activities, 16
Private Control, 26
Private property resources (PPRs), 17, 24
Private rights, 26, 33
Production-distribution chain, 17
Property rights, 3, 33

Qualitative tools, 6

Real estate sectors, 59
Reciprocal obligations, 26
Red pepper, 20
Registered architectures, 46
Registered deeds, 50
Registration Act, 33, 58
Registration fees, 48–50, 58
Regulation of Building Operations Act, 30
Regulatory mechanisms, 2
Resources
economic, 1
immovable, 1
Restoration of fertility, 19
Revenue collection, 48

Ridge stretching, 8
Rotation of Crops, 19
Rural development, 3

Sal forest, 12–13
Sample survey, 14–16
Settlement pattern, 13
Sex ratio, 15
Share tenancy, 4
Shivalik hills, 7–8
Skill development, 22, 56
Skilled labour, 43
Small scale industries, 16, 30
Snow-bound areas, 8
Snow-clad peaks, 7
Social organizations, 3
Social poverty, 55
Social practices, 2
Social sustenance, 17
Socio-economic
  activities, 1
  benefit, 53
  groups, 5
Stamp duties, 48–50, 52, 58
Steep slopes, 7
Sub-Arctic Region, 9
Sub-temperate region, 9
Sub-Tropical Forest Region, 12
Supplementary supply, 24
Surplus funds, 37
Survey of India, 29

Temperate Forest Region, 12
Temperate region, 9, 12
Terrace cultivation, 18
Terraced fields, 7, 18
Tertiary sectors, 30
Theoretical orientation, 6
Three-dimensional resources, 1
Tract
  distinct, 7
  mountain, 7, 16
  sub-mountain, 7
Traditional community institutions, 28
Transfer of land, 4, 56
Transparent system, 52
Treeless slopes, 12

Unconditional rights, 4
Undeclared money, 50
Upland slopes, 18
Urban centre, 30
Urbanized economy, 16
Uttarakhand Land Administration system, 51

Valley
  region, 17–19, 30, 52
  segment, 16
  upper, 12
Village agriculture, 47
Vote Bank, 52

Waste land, 23, 52, 73, 77, 83
Wheat, 19
Wild animals, 13, 24
Wild bees, 21

Yamuna and Ganges rivers, 8

